HEALTH EFFECTS OF DRINKING WATER TREATMENT TECHNOLOGIES

DRINKING WATER
HEALTH EFFECTS TASK FORCE

 LEWIS PUBLISHERS

Library of Congress Cataloging in Publication Data

Comparative health effects assessment of drinking water treatment technologies.
    Health effects of drinking water treatment technologies.

    Reprint, with new introd. Originally published: Comparative health effects assessment of drinking water treatment technologies. Washington, D.C.: Office of Drinking Water, U.S. Environmental Protection Agency, 1988.
    Includes index.
    1. Drinking water—Health aspects—United States.
    2. Drinking water—United States—Contamination.
    3. Drinking water—United States—Purification—By-products—Toxicology. I. Drinking Water Health Effects Task Force. II. United States. Environmental Protection Agency. Office of Drinking Water. III. Title.

    RA592.A1C64   1989   363.6'1   89-2707
    ISBN 0-87371-223-4

This previously unpublished book was prepared by a task force consisting of a panel of experts in the drinking water technology and toxicology fields working with the Office of Drinking Water, U.S. Environmental Protection Agency, Washington, D.C.

First printing, 1989, by Lewis Publishers, Inc., 121 South Main Street, Chelsea, MI 48118.

PRINTED IN THE UNITED STATES OF AMERICA

Table of Contents

Drinking Water Health Effects Task Force

Edward J. Calabrese, Ph.D.
  Chairman
Division of Public Health
University of Massachusetts
Pleasant Street, Room N340
Amherst, MA 01003

Joseph Borzelleca, Ph.D.
Medical College of Virginia, VCU
Smith Building, Room 660
12th and Clay Streets
Richmond, VA 23219

David Brown, Ph.D.
Department of Pharmacology and
  Toxicology
Northeastern University
312 Mugar Life Sciences Building
360 Huntington Avenue
Boston, MA 02115

Richard Bull, Ph.D.
College of Pharmacy
Washington State University
Wegner Hall, Room 105
Pullman, Washington 99164-6510

William D. Burrows, Ph.D.
Schaub, Burrows & Associates, Ltd.
6708 Autumn View
Eldersberg, MD 21784

Arthur Furst, Ph.D.
3736 La Calle Court
Palo Alto, CA 94306

Chuck Gerba, Ph.D.
University of Arizona
Department of Microbiology and
  Immunology
Building 90
Tucson, AZ 85721

Steven Schaub, Ph.D.
Schaub, Burrows & Associates, Ltd.
6708 Autumn View
Eldersberg, MD 21784

Ed Singley, Ph.D.
1020 N.W. 23rd Avenue
Suite D
Gainesville, FL 32609

Vern Snoeyink, Ph.D.
University of Illinois
Newark Civil Eng. Laboratory MC-
  250
208 North Romine Street
Urbana, IL 61801-2397

Rhodes Trussell, Ph.D.
250 N. Madison
Box 7009
Pasadena, CA 91101

Robert Tardiff, Ph.D.
Environ Corporation
1000 Potomac Street, NW
Washington, DC 20007

Preface

The Safe Drinking Water Act Amendments of 1986 require the U.S. Environmental Protection Agency (EPA) to assess the relative health risks of treated versus untreated drinking water. Under the Amendments:

> The Administrator of the Environmental Protection Agency shall conduct a comparative health effects assessment, using available data, to compare the public health effects (both positive and negative) associated with water treatment chemicals and their by-products to the public health effects associated with contaminants found in public water supplies. Not later than 18 months after the date of enactment of this Act, the Administrator shall submit a report to Congress setting forth the results of such assessment.

This provision originated primarily from a concern about the health risks associated with trihalomethanes (THMs) and other common disinfection by-products in drinking water treated with chlorine. Is society, Congress asked, merely trading off one health risk for another? What is the relative risk of illness due to pathogens versus cancer and other long-term health effects associated with treatment chemicals and their by-products?

For more than 10 years, EPA has been undertaking very major research efforts regarding the identification, chemistry, toxicology, and treatment technology of the by-products of disinfection. EPA has focused on identifying and quantifying the health risks associated with chlorinated by-products of disinfection; however, recent emphasis has been placed on identifying the by-products of ozonation and other disinfectants, and in characterizing their toxicity. EPA is in the process of developing disinfection treatment requirements (as required by the Safe Drinking Water Act), as well as comprehensive regulations to control disinfectants and disinfectant by-products. The intimate relationship between the two regulatory activities requires a comprehensive strategy that will optimize the benefits of disinfection while minimizing the potential risks from the chemical residues that it leaves. The historical essentiality and large public health benefits of the disinfection process must be carefully weighed against potential health risks before significant changes in water treatment practices are mandated.

Water treatment techniques in the United States have numerous and varied functions, including disinfection, filtration, corrosion control, taste and odor control, softening, and turbidity control. The adoption of active disinfection with chlorine in the early 1900s was instrumental in virtually eliminating typhoid fever and numerous other waterborne diseases in the United States. The addition of chlorine to drinking water has thus had an enormously positive impact on national public health, and the benefits realized from this disinfection practice are a major achievement.

Without disinfection and filtration, waterborne disease would surely spread rapidly in most public water systems served by surface water. Only the continued practice of disinfection and filtration in these systems can protect the public from a potential resurgence of these waterborne diseases. Microbiological threats to public health, however, have not disappeared. Viruses and protozoans, such as *Giardia,* are a particular concern because these organisms are more resistant to chlorine than are bacteria. In fact, the causes of waterborne disease outbreaks have shifted dramatically in recent years, from principally bacterial to viral and protozoan. Improved methods for detecting microorganisms in water have led to the discovery of additional viruses and protozoans responsible for waterborne disease outbreaks.

A legitimate public health concern has also been raised about the potentially harmful by-products of disinfection. Disinfection of drinking water that is high in organic content does form small amounts of THMs and other potentially toxic and/or carcinogenic by-products. THMs were regulated by EPA in 1979 as surrogates for a variety of by-products formed during chlorination. Since then, many public water systems have taken steps to reduce these by-products in finished water. The potential health risks to humans of these by-products appear small (based on existing data) relative to the historical risks of waterborne disease. Over the past decade, enough has been learned of the formation and nature of these by-products to raise public health questions, but toxicological evaluation of these chemicals and their interactions is still incomplete.

The major challenge for water utilities is to sustain a high level of disinfection while minimizing the possibility of forming by-products of disinfection that may be potentially harmful. Treatment processes have been modified, and many utilities have begun to use alternative disinfectants to chlorine such as chloramines, chlorine dioxide, and ozone. In some cases, such alternatives could eliminate the formation of THMs and other chlorinated by-products, but their effectiveness and potential human health effects must be fully evaluated. In any case, implementation of some of these alternatives will be hindered by their cost; small systems will have more difficulty setting up new processes than will large systems.

Drinking water contaminants such as lead, pesticides, and radon also present public health concerns. For example, with regard to lead, the main health risk is to children who are vulnerable to its neurotoxic and behavioral effects at relatively low levels. Corrosive waters in contact with lead pipe and other lead-containing material can leach significant amounts into delivery water.

In a concerted effort to reduce lead exposure and notify the public of the hazards presented by lead, EPA has mandated the reduction of lead in gasoline, proposed a marked reduction in the lead drinking water standards, and required public notification of those who may be exposed to lead in drinking water. In addition, the SDWA amendments of June 1986 banned the use of lead solders, flux, and pipes in water distribution systems to the percentages specified in Section 1417(d) of the Act.

This book evaluates the public health impact of the most widespread drinking water treatment technologies, with particular emphasis on disinfection. While there are other developing technologies such as membrane filtration, ozone/ultraviolet, and ozone/peroxide, this report focuses solely on the most common treatment technologies and practices used today.

The book was prepared by EPA, in conjunction with a group of experts in the drinking water treatment technology and toxicology fields, and has been reviewed by the EPA Science Advisory Board (SAB).

FINDINGS

A. In general, commonly used drinking water treatment processes—including disinfection, filtration, and other processes designed to remove chemical and physical contaminants—provide enormous benefits in ensuring the quality and safety of drinking water in the United States. While the available data did not allow the Agency to perform a fully quantitative comparison, the potential negative aspects of these technologies appear to be small (based on existing epidemiologic and animal data) relative to their benefits. This book focuses on the risks of disinfection and its by-products since it is the most widespread treatment technique employed on a national basis and the public is generally exposed to disinfection by-products, almost exclusively from drinking water, over a lifetime. Additional information on the subchronic and chronic effects of disinfectant by-products is needed in order to fully assess the potential human health risks of these compounds. Other treatment techniques have been addressed in this book but with less emphasis since the probability of widespread exposure to by-products of potential concern from these treatment techniques is usually much lower than for disinfection and available data on their health effects are limited. EPA will continue to strive, in its regulatory program, to optimize the beneficial uses of these water purification technologies while further reducing any negative aspects that may remain.

B. Pathogenic microorganisms continue to be the major cause of waterborne disease outbreaks in the U.S. Many of these outbreaks are associated with microorganisms, especially viruses and protozoans not previously identified as having a waterborne route of transmission. Some of these microorganisms offer a substantial challenge to treatment or removal because of their resistance to disinfection or because existing treatment trains in some public water systems are not optimized to remove or inactivate them.

C. Raw water quality is highly variable in its content of microbes, organic substances, inorganic compounds, and radionuclides. Since pathogens clearly exist in many raw waters, disinfection and/or other technologies designed to remove them should be used by those water systems with contaminated water to ensure protection against significant public health risks. In addition, chemical contaminants both natural and synthetic (e.g., from various industries) may be present in drinking water and can be managed, through various treatment technologies, to avoid adverse health risks, which depend in part upon the concentrations of the chemicals in water.

D. To date, analytical surveys have discovered a number of organic by-products of chlorination. Among the most common are chloroform, dichloroacetic and trichloroacetic acids, trichloroacetaldehyde, bromodichloromethane, dibromochloromethane, 1,1,1-trichloropropanone, and dichloroacetonitrile. Based on the existing data, several of these compounds may be of human health concern at certain levels of exposure.

E. The adverse health effects and risks from pathogens resulting from a lack of disinfection are based on numerous human epidemiologic data and are well known. In contrast, the risks from disinfected drinking water are not as well defined due to lack of toxicity data on many of the chlorination by-products. However, given the existing epidemiologic data on chlorinated water, the magnitude of these risks is likely to be smaller, in general, compared to the risks of drinking water that is not disinfected.

F. Other disinfectants have been less extensively studied than chlorine. Existing data would indicate that, as with chlorine, organic compounds of potential toxicological significance are formed to some degree by the addition of ozone, chlorine dioxide, or chloramines to drinking water. As part of the ongoing regulatory process, EPA is examining the health effects of the by-products of these alternative disinfectants.

G. On the other hand, chlorine, chlorine dioxide, and monochloramines themselves as well as the inorganic by-products of chlorine dioxide show some evidence of toxicity at high doses. The significance of these effects in humans at low doses needs further evaluation. The Agency is currently conducting studies on the toxicology of these disinfectants to assist in the development of drinking water standards.

H. Over the past decade, several hundred outbreaks of waterborne disease have affected more than 100,000 persons. These outbreaks have occurred primarily in areas served by systems where adequate treatment, including disinfection, was not applied or was not maintained.

I. Since EPA published the 1979 THM regulation, a striking shift has occurred in disinfection methods, significantly reducing exposure to chlorination by-products. The number of communities using chloramines for disinfection has increased from less than 3% to more than 25% in large surface water systems and 13% in small surface systems. Most of these are using chloramines to maintain a disinfectant residual in the distribution system following treatment—a common procedure to ensure that drinking water remains disinfected during its journey to the tap. Considerable interest has developed concerning the use of other disinfectants as an alternative to chlorine in order to reduce THM formation. However, a community's choice of disinfectant is often affected by cost, and chlorine is often more economical.

J. Another health concern is the low-level contamination of drinking water by industrial compounds. These include the common solvents, such as trichloroethylene (TCE) and tetrachloroethylene (PCE), as well as pesticides in agricultural areas. Current treatment technologies, such as coagulation, sedimentation, and filtration, can remove a number of synthetic organic chemicals (SOCs) but, in general, removal efficiencies range from 30% to 98% and are not consistently high. Granular activated carbon (GAC) and air stripping are effective treatment technologies for organics, especially volatile organic compounds (VOCs). Air stripping is not effective in removing nonvolatile halogenated chemicals, which may be of more concern than VOCs.

K. Pipe and/or solder corrosion, which is more of a problem in some regions of the country than in others, releases several metals into drinking water, notably lead. Lead levels in drinking water are of particular concern for children, who are vulnerable to its neurotoxic and behavioral effects at relatively low exposure levels. The SDWA banned the installation of plumbing materials containing percentages of lead higher than those specified in the Act (not more than 0.2% for solders or fluxes, and 8.0% for pipes and pipe fittings) in any building connected to a public water system after June 19, 1986. Approximately 16% of public water systems in the United States have very corrosive water, and 57% have water that is moderately corrosive. The relationship between corrosivity and lead levels in drinking water is not precisely quantified. However, additional corrosion control is needed to reduce drinking water lead levels in those cases where lead levels pose adverse health effects and are significantly affected by corrosivity of water. This has been addressed in revised standards for lead that were proposed on August 18, 1988.

L. Aluminum sulfates are among the most common coagulants and, as such, may be introduced into finished drinking water. Aluminum toxicity in drinking water is difficult to assess since humans receive much more exposure to aluminum from other sources. Existing data do not support inferences that ingested aluminum in water causes either neurotoxicity or Alzheimer's disease in humans.

M. Sodium can be introduced into drinking water during the water softening process, and it is commonly present from natural sources. The role of drinking water sodium in the etiology of human hypertension remains a matter of controversy. However, elevated levels of sodium in drinking water may be an important concern for individuals with genetic predisposition to hypertension and who are required to have highly restricted intake.

N. Drinking water drawn from ground water is normally a small (less than 5%) contributor to indoor air radon. Even so, the possible health risk from radon in water is a concern. In one analysis, EPA calculated that exposure to natural radon volatilized from drinking water could result in up to 30 to 600 excess lung cancer deaths per year in the United States.

O. Asbestos can be introduced into drinking water either from natural sources or as a result of corrosion of asbestos-cement pipe. Data indicate that asbestos is carcinogenic in humans through inhalation exposure. However, most of the data regarding ingestion risks have been negative. EPA has proposed a Maximum Contaminant Level Goal for fibers longer than 10 micrometers to conservatively protect against the possibility that they may be carcinogenic through ingestion. Cancer risks, if any, should be minimal under current exposure levels.

P. The environmental or public health effects from managing the disposal of filtration backwash, coagulant sludges, exhausts from air stripping of VOCs and radon, and the regeneration and disposal of granular activated carbon have not been fully investigated or quantified in this book. Preliminary data on air stripping indicate that ambient levels of VOCs and radon will probably not be large enough to pose a significant health risk. The health risks associated with uncontrolled regeneration and disposal of contaminated GAC may be of concern but have not been completely addressed in the scientific literature. The Agency is presently studying methods for the proper disposal of wastes from these treatment techniques.

CONCLUSIONS

1. Since alternative disinfectants to chlorine are expected to be increasingly adopted in the United States, high priority is being given to further

characterizing the chemical and toxicological nature of their by-products, particularly those of ozone.

2. Because chloramine is a significantly weaker disinfectant than chlorine, preformed chloramination should not be used as a method of primary disinfection, especially when viruses and/or parasites are a concern, but its use as a secondary disinfectant is acceptable. The use of monochloramine should be explored, especially with regard to its effectiveness in controlling disease-causing bacteria, such as *Legionella,* during water transport.

3. When free chlorine (chlorine that has not reacted with other compounds) is employed as the primary disinfectant and the potential for THM formation is significant, water utilities should maintain a slight residual of free chlorine above that needed to oxidize nitrogen and then add ammonia. This technique results in monochloramine formation and reduces the potential for forming THM and other by-products.

4. An effort should be made to remove significant levels of organic precursors prior to disinfection. This improves the efficiency of disinfection, reduces the formation of disinfection by-products, and may obviate the need for additional treatment to remove excess by-products.

5. Corrosion control is important in many public and private water systems given (1) the health concerns associated with corrosion by-products (especially lead), (2) its enormous impact on distribution system replacement, and (3) the effect of corrosion on the esthetic quality of water.

6. GAC or air stripping is highly recommended where high removal rates are necessary to meet standards for volatile organic compounds. Air stripping is also known to be effective in removing radon.

Introduction

Public concern with both the quality and safety of drinking water has likely stemmed from:

- The widespread publicity concerning the contamination of ground water with synthetic organic chemicals.
- Pesticide infiltration of some drinking water supplies.
- The presence of chloroform and other by-products in drinking water as a result of the disinfection of surface water by chlorine.
- Elevated levels of sodium in drinking water from natural sources as well as municipal de-icing operations and water treatment.
- Leaching of lead from old lead pipes and from lead solder in newer pipes.
- The emerging public recognition of radon in numerous ground-water sources in many parts of the country.

Mechanical treatment (e.g., filtration) and disinfection of drinking water have dramatically reduced waterborne diseases such as typhoid fever, cholera, hepatitis, and other enteric virus infections in the U.S. population. A classic example is the marked decrease in typhoid fever. In 1900, the annual typhoid death rate in the United States was 36 per 100,000 people. As water treatment (chlorination) became more common, the rate declined to 20 per 100,000 in 1910, to 3 per 100,000 in 1935 (Akin, 1982), and to virtually zero (water-related) today. However, newly discovered (but not new) viral, bacterial, and protozoan pathogens in water require even better procedures of treatment and disinfection. The conventional disinfection and filtration techniques commonly practiced also provide significant protection from these microorganisms.

With widespread chlorination, however, came chlorination by-products such as chloroform, chloroacetic acids, and chlorinated phenols that may pose some cancer and other health risks to the general population. To address this issue, Congress included in the Safe Drinking Water Act Amendments of 1986 a provision requiring the U.S. Environmental Protection Agency (EPA) to assess and compare the health effects associated with water treatment chemicals and their by-products to contaminants found in untreated drinking water supplies. Specifically, the provision states that:

> The Administrator of the Environmental Protection Agency shall conduct a comparative health effects assessment, using available data, to compare the public health effects (both positive and negative) associated with water treatment chemicals and their by-products to the public health effects associated with contaminants found in public water supplies. Not later than 18 months after the date of enactment of this Act (June 19, 1986), the Administrator shall submit a report to Congress setting forth the results of such assessment.

The legislative history of this provision provides interesting insight into its origin. Apparently, this provision stems primarily from a Congressional concern over the health risks associated with trihalomethanes (THMs), a common disinfection by-product in finished drinking water. Is society, Congress asked, merely trading off one health risk for another? For example, are we trading off sickness and death due to pathogens for cancer and other long-term effects? Consequently, this book focuses on disinfection since this is of greatest public concern and is the most widespread treatment technique employed on a national basis. To a lesser degree, the book also examines the health risks from the other major treatment technologies. Developing technologies, such as membrane filtration, ozone/ultraviolet, and ozone peroxide, are not addressed.

In response to the Congressional charge, the Office of Drinking Water assembled a committee of experts in drinking water treatment technology and toxicology. These experts were directed to evaluate, as far as the available data allow, the relative benefits and risks of each of the common types of drinking water treatment technologies. This generally involved an assessment and comparison of the risks associated with:

- Nontreatment (e.g., from pathogens in the case of disinfection or chemicals in the case of corrosion).
- The actual chemicals used in treatment.
- The by-products of those chemicals.

Quantitative assessments are not always possible because of a lack of toxicity and human exposure data concerning a particular treatment type or chemical occurrence. When quantitative data were lacking, EPA directed each toxicologist to make the best assessment possible of the potential health effects for a specific technology.

The book also addresses the challenges that water treatment systems face when the removal of one contaminant leads to the formation of others or when optimizing treatment to control one substance causes others to increase (e.g., low pH reduces THMs but increases corrosion).

This book is intended for both the layperson and the technical reader. It should serve as a:

- Guide for the regulatory community for identifying areas where problems exist and how they may be addressed.
- Guide for the general public on the quality of the country's drinking water, by putting drinking water risks in a broader public health perspective.
- State-of-the-art judgment of which treatment technologies have worked well and thus should be sustained or, where feasible, improved.

The book is divided into seven chapters. Chapter 1 assesses raw water quality, and Chapter 2 provides an overview of drinking water treatment, which usually involves various treatments working in concert rather than a single technology. This chain of treatments, known as a process train, makes risk analysis complex because many possible by-products can be generated and exposure levels can be expected to fluctuate. Chapters 3 through 7 present the treatment technologies and their associated health risks (from substances either added or removed due to treatment). These chapters are organized according to treatment type, since the majority of chemicals of concern are treatment chemicals or by-products of treatment. In essence, the treatments served as a focal point for analyzing the risks associated with the introduction and removal of chemicals from drinking water.

REFERENCE

Akin, E.W. 1982. Waterborne outbreak control: which disinfectant? *Environ. Hlth. Perspect.* 46:7–12.

1

Raw Water Quality

SUMMARY

This chapter discusses the microbiological and chemical constituents of raw water and their implications for human exposure. Raw water sources consist of surface waters and ground waters. Deep protected aquifers are the preferred source of potable water because years of slow filtration have left the water virtually free of turbidity, organic matter, and pathogens. Pathogens, rather than chemicals, have historically been the major health concern in drinking water.

Pathogenic microorganisms clearly exist in most raw source waters, especially surface waters. To protect the public's health, they must be reduced to safe levels; i.e., levels that protect the public from infectious outbreaks. Most drinking water problems are of microbiological origin and are caused by inadequate or improper treatment; however, the incidence of waterborne disease in the United States has been significantly lowered as a result of adequate treatment and disinfection.

The general properties and chemical constituents of water affect the prevalence of microorganisms. Important general properties include pH; alkalinity, a measure of acid-buffering strength; hardness, a measure of the level of soluble calcium and magnesium; and turbidity, a characteristic of primary importance that reflects the amount of suspended soil and sediments in water.

The inorganic chemicals in drinking water include common salts, heavy metals, asbestos, fluoride, nitrate, and radionuclides. Some heavy metals that may reach significant concentrations in public water supplies are arsenic, copper, iron, lead, and selenium.

Organic chemicals found in significant concentrations in water include both natural and synthetic chemicals. On a quantitative basis, natural organic chemicals are predominant. They come from soil run-off, forest canopy drip, aquatic biota, and human and animal wastes. Synthetic organic chemicals include pesticides, herbicides, and chemicals that originate from industrial or human activities. They sometimes enter water supplies from run-off, leachate from waste disposal areas, precipitation from the atmosphere, or discharge from sewage treatment plants.

The health implications of chemical contaminants depend on their toxicological characteristics and their concentrations in water.

1.1 Introduction

Raw water sources are commonly broken down into surface waters and ground waters. Surface waters include rivers, lakes, and reservoirs fed by direct rainfall and run-off, whereas ground waters include deep and shallow aquifers and springs, fed by the slow percolation of surface waters through soil and subsoil layers.

In general, surface waters have rapid replacement rates, from virtually instantaneous in the case of rivers, to hundreds of years for large, deep lakes. Replacement rates for ground waters, on the other hand, are relatively slow—thousands of years in some cases. Contaminated surface waters tend to purify themselves once the source of contamination is removed; ground waters tend to stay contaminated because the water moves slowly and the source of contamination is difficult to remove. The technology for cleaning contaminated aquifers is just now becoming available, and it is costly. Generally, where deep, protected aquifers are available, they are a preferred source of potable water, because years of slow filtration have left the water free of turbidity, synthetic organic matter, and pathogens.

Table 1-1 shows typical data from the Centers for Disease Control (CDC) reports on waterborne outbreaks in the United States. It should be noted that not all waterborne disease outbreaks are reported. The 1985 CDC report cautions that these data "should not be used to draw firm conclusions about the true incidences of waterborne disease outbreaks, the seasonality of outbreaks, and the deficiencies in water systems that most frequently result in outbreaks." Despite this comment, the 1985 CDC report, as well as reports from 1977 through 1983, show that most clearly documented drinking water problems are of microbiological origin. Other problems, such as chronic health effects, are not as easily documented.

Table 1-1. Reported Waterborne Outbreaks, United States, 1984[a]

State	Month	Etiology[b]	Cases	Type of System[c]	Deficiency[d]	Location of Outbreak	Location of Source
AK	Sep	*Giardia*	3	I	2	camp	stream
AK	Oct	*Giardia*	123	C	3	community	reservoir
CA	Nov	copper	1	I	4	high school	soda machine
CO	Nov	*Giardia*	13	C	3	community	river
CO	Mar	*Giardia*	400	NC	3	ski resort	pond
CO	Aug	AGI	50	C	3	community	river
ID	Mar	*Campylobacter*	6	C	3	community	spring
MA	Sep	Hepatitis A	7	I	2	household	well
MN	Mar	*Campylobacter*	9	I	2	household	well
MO	Oct	AGI	107	C	4	airport restaurant	sewage overflow
MO	Jun	*Entamoeba*	4	C	2	trailer park	well
MO	Jun	AGI	2	I	2	household	well
NC	Feb	copper	1	I	5	workshed	well
NY	Jun	*Campylobacter*	4	NC	2	amusement park	spring
OR	Jun	*Campylobacter*	22	C	3	community	wells, creek
OR	Jul	*Giardia*	42	C	3	community	river
OR	Aug	AGI	20	C	4	plywood mills	river
PA	Feb	*Giardia*	298	C	3	community	river
PA	May	AGI	8	I	2	picnic	well
PA	Aug	AGI	98	NC	2	resort	well
PA	Sep	AGI	34	I	2	bicycle race	private well
PA	Oct	AGI	18	I	2	industry	well
TX	May	Norwalk agent	251	C	3	community	well
TX	Jul	*Cryptosporidium*	117	C	3	community	well
WI	Mar	AGI	89	NC	4	restaurant	sewage overflow
VA	Mar	crude oil	28	C	5	community	spring

[a]Adapted from Centers for Disease Control (1985).
[b]AGI = acute gastrointestinal illness of unknown etiology.
[c]C = community (municipal); NC = noncommunity (semi-public); I = individual.
[d]1 = untreated surface water; 2 = untreated ground water; 3 = treatment deficiencies; 4 = distribution system deficiencies; 5 = miscellaneous.

1.2 Microorganisms and Disease

Microorganisms are the major cause of illness associated with the food and water we consume. This illness is costly: the worldwide impact of food-borne illnesses associated with microorganisms is estimated at $20 billion annually (Todd, 1987).

On an international scale, an estimated 25,000 persons per day died in 1980 from consumption of contaminated water, and one hospital bed in four was occupied by someone who became ill from polluted water (WHO,

1979). Perhaps 80 percent of all diseases in the world are related to contaminated water (WHO, 1979). An estimated 5 billion waterborne infections occur annually in Africa, Asia, and Latin America (Walsh and Warren, 1979). Microbiological contamination of food and water is one of the major causes of diarrhea, resulting in up to 1 billion cases every year in children less than 5 years old (WHO, 1984). Diarrhea is a frequent cause of mortality in most of the developing world today (MMWR, 1983).

These statistics largely concern the developing world where, compared to the United States and the rest of the industrialized world, both the quality and quantity of drinking water are diminished. The incidence of disease caused by waterborne microorganisms is low in the United States due to advances in both wastewater and drinking water treatment. However, disease-causing microorganisms are still present in natural waters that are used as raw sources for our drinking water supplies—a reality that must not be ignored. Without disinfection and other treatment technologies to reduce or eliminate microorganisms, human health (irrespective of medical care and nutritional status) would be seriously jeopardized.

1.2.1 Causes of Waterborne Disease

Most raw surface waters and many ground-water sources in the United States have microbial populations composed of bacteria, viruses, protozoa, fungi, helminths, and algae. The mere presence of microorganisms does not necessarily mean that water is unsafe to drink; in fact, many common organisms in water are not pathogens and are beneficial because they help decompose organic nutrients and detritus that enter the water from natural and manmade sources. However, failure to adequately and continuously remove harmful microorganisms is a continuing cause of waterborne disease in the United States today (Craun, 1986). Most disease outbreaks are believed to be due to the use of untreated water, inadequately treated water, or water contaminated after treatment.

Figure 1-1 shows the causes of waterborne disease outbreaks recorded in a 35-year retrospective study. Over half the outbreaks were due to the use of inadequately treated or untreated water. Outbreaks occurred from the use of both surface and ground water. In another study that examined waterborne disease over a six-year period, approximately 23 percent of the outbreaks were due to contaminated surface water while 56 percent were due to contaminated ground water (Figure 1-2). Disinfection is more commonly used for surface water supplies than for ground-water supplies (e.g., small wells). This explains the larger number of reported outbreaks due to contaminated ground water.

Inadequate or interrupted treatment and distribution problems often lead to waterborne disease, regardless of the source (Craun, 1986 cited in

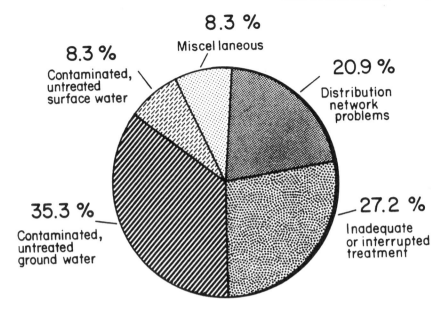

8.3 %
Miscellaneous

8.3 %
Contaminated,
untreated
surface water

20.9 %
Distribution
network
problems

35.3 %
Contaminated,
untreated
ground water

27.2 %
Inadequate
or interrupted
treatment

Figure 1-1. Waterborne disease outbreaks caused by deficiency in public water systems. *Source:* Lippy and Waltrip, 1984. Reprinted from *American Waterworks Association Journal,* Vol. 76, No. 2. (February 1984), by permission. Copyright 1984, American Waterworks Association.

Geldreich, 1986) (Table 1-2). From 1920 to 1983, 1,531 waterborne disease outbreaks were reported in the United States (Craun, 1986). A microbial pathogen could be identified as a probable cause in 48 percent of the outbreaks and chemicals in 4 percent. Outbreaks of unknown cause that resulted in illness were categorized as acute gastroenteritis (Craun, 1986). It is difficult to estimate how many of these outbreaks were caused by microorganisms because either clinical specimens were not collected or the causative organism could not be isolated or identified.

As methods for detecting and identifying parasites and viruses have become available, the number of outbreaks attributed to these organisms has increased (Craun, 1986; Rose and Gerba, 1986). Over the last decade and a half, an average of two new enteric viruses that could be transmitted through potable water have been discovered each year (Gerba, 1984). This list of potential waterborne disease organisms is expected to continue to grow; however, many laboratories responsible for water analysis lack the technology, resources, and/or personnel to detect or identify these new microbial types. The complexity and slowness of analysis of biological agents are major reasons why public health officials have historically relied upon the installation and operation of appropriate water treatment technologies—

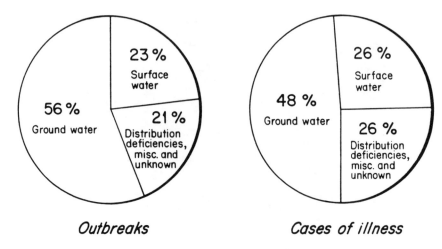

Figure 1-2. Waterborne disease outbreaks in the United States (1971–1977). *Source:* Craun, 1979.

Table 1-2.   Water System Deficiencies Causing Outbreaks of Diseases, 1920-1983[a]

Public Water System Deficiencies	No. of Systems	% Deficiencies	No. of Cases of Illness
Contaminated, untreated ground water	661	43.2%	82,528
Inadequate or interrupted treatment	333	21.8%	224,973
Distribution network problems	233	15.2%	83,577
Contaminated, untreated surface water	158	10.3%	12,709
Miscellaneous causes or insufficient evidence	146	9.5%	11,542
Total	1,531	100.0%	415,329

[a]Data adapted from Craun (1986) cited in Geldreich (1986).

rather than detailed water monitoring—to ensure the safety of the finished drinking water.

1.2.2 Microorganisms Associated with Waterborne Disease

Water-related disease organisms have been classified by White and Bradley (1972) as:

- *Waterborne,* which are pathogens that originate in fecal material and are transmitted by drinking water.
- *Water-washed,* which are organisms that originate in feces and are transmitted through contact because of inadequate sanitation or hygiene.
- *Water-based,* which are organisms that spend part of their life cycle in aquatic animals and come in direct contact with humans in water, often through the skin; and
- *Water-related,* which are microorganisms with life cycles associated with insects that live or breed in water and bite susceptible individuals.

In particular, many of the waterborne, water-washed, and water-based organisms can be present in raw water (see Table 1-3).

Table 1-3.   Water-Related Diseases and Route of Transmission[a]

| | Route of Transmission | | |
Water-related Diseases (worldwide)	Water-borne	Water-washed	Water-based
Bacterial diseases			
Bacillary dysentery (*Shigella* spp.)	X	X	–
Cholera (*Vibrio cholera*)	X	–	–
Diarrhea (*Campylobacter*)	X	–	–
Diarrhea (*Eschericha coli*)	X	X	–
Leptospirosis (*Leptospira* spp.)	X	X	–
Salmonellosis (*Salmonella* spp.)	X	X	–
Typhoid fever (*Salmonella typhi*)	X	X	–
Skin infections (*Pseudomonas* spp. and *Staphylococcus* spp.)	–	X	–
Yersiniosis (*Yersinia* spp.)	X	–	–
Viral diseases			
Enteroviruses	X	X	–
Gastroenteritis, Norwalk agent and Rotavirus	X	–	–
Hepatitis A (*Hepatitis virus*)	X	X	–
Parasitic diseases			
Acanthamebiasis (*Acanthamoeba* spp.)	X	–	–
Amoebic dysentery (*Entamoeba histolytica*)	X	–	–
Ascariasis (*Ascaris lumbricoides*)	X	X	–
Balantidial dysentery (*Balantidium coli*)	X	X	–
Dracontiasis (*Dracunculus medinensis*)	–	–	X
Giardiasis (*Giardia lamblia*)	X	–	–
Meningoencephalitis (*Naegaleria* spp. & *Acanthamoeba* spp.)	X	–	–
Schistosomiasis (*Schistosoma* spp.)	–	–	X

[a]Cooper et al. (1986).

Enteric pathogens typically associated with waterborne disease in the United States are presented in Table 1-4. These pathogens typically include bacteria, viruses, and parasites (protozoans and helminths) that are excreted in the feces of humans and animals. Once excreted, these organisms may survive for prolonged periods of time in water, even months to years under the proper environmental conditions (Feachem et al., 1983). Many pathogens, particularly viruses and parasites, can survive or escape conventional sewage treatment in concentrations capable of causing disease (Feachem et al., 1983). Therefore, these organisms may constitute a potential source of raw water contamination.

Currently, *Giardia* occurs widely in the United States and is the most commonly identified organism associated with waterborne disease outbreaks in this country (Craun, 1986). Other currently documented agents of waterborne diseases in the U.S. include bacterial species of *Shigella, Salmonella, Campylobacter,* and *Yersinia:* hepatitis, Norwalk, and rota viruses; and protozoans of the genus *Cryptosporidium.* Additional concern for human health arises from opportunistic microorganisms of nonfecal origin including the bacterial genera, *Legionella, Mycobacterium, Pseudomonas,* and various fungi.

1.2.3 Nature and Fate of Pathogens in Water

The fate of microorganisms in water varies. A number of factors in water affect the condition of the organisms and their ability to reach and infect susceptible human hosts. One of the most important factors is pH. Generally, microbial survival is prolonged at a pH near neutrality (pH = approximately 7). When the pH is less than 6 or exceeds 8, typical enteric pathogens often show an increased rate of die-off or inactivation; however, not all pathogens react to pH variations in the same way. Temperature also plays a significant role. Usually water temperatures near freezing prolong survival, whereas temperatures of about 25°C or higher reduce survival. However, a few gastrointestinal tract bacterial pathogens can actively metabolize and reproduce at elevated temperatures in water. The salinity of water may affect survival, but no firm trends exist for the majority of waterborne pathogens.

Most surface waters and some ground waters contain particulate or colloidal materials from human pollution (sewage, industrial wastes, agricultural run-off, real estate/commercial development) and from natural processes of weathering and meteorological events. Microorganisms, especially bacteria and viruses, become adsorbed to or enmeshed in these materials, travel with them, and settle out in conjunction with them, especially in the presence of dissolved salts which enhance the particulate association. If the solids permanently settle to the bottom of a water source, such as a lake, the microorganisms can be effectively removed. However, in streams, rivers, and small

Table 1-4. Summary of Taxonomic, Clinical, and Epidemiological Features of Potential Drinking Water Disease Agents

Name of Organism or Group	No. Types	Major Disease	Major Reservoirs/Primary Sources	Conc. in Primary Source	Infect. Dose	Prevalence (Avg. % Excretion)	Duration of Shedding	Carrier* State
Bacteria								
Salmonella typhi	1	typhoid fever	human feces	10^6/g	low	[	[typically	+
Salmonella paratyphi	1	paratyphoid fever	human feces	10^6/g	high	[~0.1]	[4 weeks.]	
Other *Salmonella* spp.	>2,000	salmonellosis	human/animal feces	10^6/g	high	[	[occ. >1 year]	+
Shigella spp.	4	bacillary dysentary	human feces	10^6/g	medium		±1 week	+(?)
Vibrio cholera	?	cholera	human feces	10^6/g	high		?	?
Enteropathogenic *E. coli*	?	gastroenteritis	human feces	10^8/g	high	1.2-15.5	?	+
Yersinia enterocolitica	?	gastroenteritis	human/animal feces	?	high	?	?	?
Compylobacter jejuni	>1,000	gastroenteritis	human/animal (?) feces	?	?	?	1-3 weeks	?
Legionella pneumophila and related bacteria	7	acute respiratory illness (legionellosis)	thermally enriched waters	?	high?	?	?	?
Mycobacterium tuberculosis	1	tuberculosis	human resp. exudates	?	?	?	?	?
Other (atypical) mycobacteria	≥2	pulmonary illness	soil and water	?	?	?	?	?
Opportunistic bacteria	?	variable	natural waters	?	?	?	?	?

Table 1-4. (continued)

Name of Organism or Group	No. Types	Major Disease	Major Reservoirs/ Primary Sources	Conc. In Primary Source	Infect. Dose	Prevalence (Avg. % Excretion)	Duration of Shedding	Carrier* State
Enteric Viruses								
Enteroviruses								
Polioviruses	3	poliomyelitis	human feces					
Coxsackieviruses A	23	aseptic meningitis	human feces					
Coxsackieviruses B	6	aseptic meningitis	human feces	10^6/g		[~1.5%]		
Echoviruses	31	aseptic meningitis	human feces		low		1-3 weeks	
Other enteroviruses	≥4	encephalitis	human feces					
Reoviruses	3	mild UR and GI ill.	human/animal feces	10g part/g		?	1-2 weeks	
Rotaviruses	≥3	gastroenteritis	human feces			?	1 week	?
Adenoviruses	37	UR and GI illness	human feces	10^6/g		?	?	
Hepatitis A virus	1	Infectious hepatitis	human feces			?	±3 weeks	
Norwalk and related GI viruses	≥3	gastrointeritis	human feces			?	≤1 week?	
Protozoans								
Acanthamoeba castellani	1	amebic meningoencephalitis	soil and water	?	?	very low	none	+
Balantidium coli	1	balantidiasis (dysentery)	human feces	?	?	very low	variable	?
Entamoeba histolytica	1	amebic dysentery	human feces	10^5/g	low	3-10	variable	+
Giardia lamblia	1	giardiasis (gastroenteritis)	human & animal feces	10^5/g	low	1.5-2.0	6-7 weeks	+
Naegleria flowieri		primary amebic meningoencephalitis	soil and water	?	?	very low	none	(months to years)

Table 1-4. (continued)

Name of Organism or Group	No. Types	Major Disease	Major Reservoirs/ Primary Sources	Conc. In Primary Source	Infect. Dose	Prevalence (Avg. % Excretion)	Duration of Shedding	Carrier* State
Helminths								
Nematodes (roundworms)								
Ascaris lumbricoides	1	ascariasis	human (& animal) feces	$10\text{-}10^4$/g	—	<1	[usually	+
Trichuris trichiura	1	trichuriasis	human feces	$10\text{-}10^4$/g	—	<1	months,	+
Hookworms							occas.]	
Ancylostoma duodenale	1	hookworm disease	human feces	$10\text{-}10^4$/g	low	<1		+
Necator americanus	1	hookworm disease	human feces	$10\text{-}10^4$/g	—	<1	[years]	+
Strongyloides stercoralis	1	threadworm disease	human & animal feces	$10\text{-}10^4$/g	—	<1		+
Algae (blue-green)								
Anaboena flas-aquae	6**+	gastroenteritis	natural waters	$10^4\text{-}10^{-6}$	?	—	—	—
Microcystis aeruginosa	4	gastroenteritis	natural waters	$10^4\text{-}10^{-6}$	?	—	—	—
Aphanizomenon flas-aguae	1+	gastroenteritis	natural waters	$10^4\text{-}10^{-6}$	?	—	—	—
Schizothrix calcicola	1	gastroenteritis	natural waters	$10^4\text{-}10^{-6}$	?	—	—	—

Source: Sobsey, M.D. and B. Olson. 1983. Microbial agents of waterborne disease. *In* Berger, P. S. and Y. Argamon (eds.), Assessment of microbiology and turbidity standards for drinking water. USEPA Report No. 570/9-83-001. USEPA, Washington, D.C.
*Untreated subclinical and/or asymptomatic infections.
**Refers to number of different strain types.
+ Also includes those whose role in disease is uncertain.
 UR = Urinary/Respiratory.
 GI = Gastrointestinal.

impoundments, this settling process may be only temporary: changing flow characteristics can remobilize the pathogen-containing particulates into the water phase.

A pathogen's association with turbidity may be either detrimental or beneficial to its survival, depending on the nature of the association. Organisms associated with turbidity can be readily filtered out by water treatment technology. On the other hand, if water is filtered ineffectively or not at all, turbidity (especially of organic composition) may protect organisms from chemical disinfectants, thus allowing the pathogens to survive.

Some natural waterborne microorganisms, their toxins, and microbial predators can have a considerable impact on the survival of pathogens in raw water. An understanding of all of the factors that contribute to the microbial and physical makeup of raw water is essential to making decisions about appropriate treatment. These factors are discussed in subsequent chapters.

1.2.4 Infectious Doses of Microorganisms

The presence in water of microorganisms that are pathogenic for humans is worrisome, but need not constitute a major health problem, provided that the numbers present are below the level that constitutes an infectious dose. For each species or strain of pathogenic virus, bacteria, protozoa, or helminth, there is a minimum level that can be expected to cause infection in a susceptible human. Usually, hundreds or thousands of viable organisms are needed to establish a bacterial infection, whereas viruses (Ward and Akin, 1984) and protozoans (Rendtorff, 1954) can establish infection with as few as 1 to perhaps 100 viable organisms (Sobsey and Olsen, 1983).

Human infection can often be established even though too few organisms are present to induce overt clinical symptoms. With or without clinical symptoms, persons infected with microorganisms in the gastrointestinal tract can transmit disease to others because significant numbers of the organisms are excreted in feces. Because of the significant risk of this type of secondary infection, high standards of microbial control must be maintained.

1.2.5 Occurrence of Pathogens in Untreated Water

Although consumption of water containing enteric disease microorganisms is clearly a significant cause of human disease, the extent of the problem is not well defined even in the United States. Many disease-causing organisms are difficult or costly to detect and identify, or cannot be detected by currently available technologies. The estimated concentrations of enteric pathogens in polluted surface waters in the United States are shown in Table 1-5. The actual concentrations probably vary greatly, depending on the incidence

Table 1-5. Estimated Levels of Enteric Organisms in Sewage and Polluted Surface Water in the United States

Organism	Concentration (per 100 mL) in	
	Raw Sewage	Polluted Stream Water
Coliforms	10^9	10^5
Enteric viruses	10^2	$1-10$[a]
Giardia	10	$0.1-1$[b]
Cryptosporidium	$10-10^3$	$0.1-10^2$[c]

[a]EPA (1978).
[b]Jakubowski (1984).
[c]Rose et al. (1987)

of disease in a community or animal population, time of year, sanitation within the community, treatment of contaminated sources, and environmental factors.

Because of the difficulty of isolating and identifying enteric pathogens in water, indicator bacteria have been used for over 70 years to judge the sanitary quality and level of fecal contamination. Total coliform, fecal coliform, fecal streptococci, E. coli, and heterotrophic bacteria are common inhabitants of the gastrointestinal tract of all warm-blooded animals, and their presence in water is an indication of probable fecal contamination. EPA considers enterococci and E. coli indicators of fecal contamination (EPA, 1986). Like enteric disease-causing microorganisms, they are introduced to surface waters from improperly treated sewage; human waste; street, storm, and feed lot run-off; grazing lands in agricultural areas; and wildlife excreta. Analysis of surface water quality data by the U.S. Geological Survey indicates that, from 1975 to 1978, approximately 50 to 99 percent of the surface water measurements taken during this 3-year period exceeded 200 fecal coliforms/100 mL for parts or all of this 3-year period (Council on Environmental Quality, 1979). These data suggest that fecal contamination of natural surface waters is prevalent and that pathogens can be found in these drinking water sources.

Generally, ground water from deep aquifers has few microorganisms present; however, this attribute should not be taken for granted. Given the proper conditions, disease-causing microorganisms, particularly viruses, may travel long distances through soil strata (Gerba and Bitton, 1984). A few extensive bacteriological surveys of ground-water quality in the United States have been conducted and are summarized in Table 1-6. In these studies, 9 percent to 85 percent of the samples examined contained coliforms, and 2 percent to 75 percent were positive for fecal coliforms. Enteric viruses commonly occur in ground waters subject to contamination by do-

Table 1-6. Microbial Surveys of Drinking Water Wells

Survey	No. of Samples	Percent Positive for Coliforms	Percent Positive for Fecal Coliforms	Reference
South Carolina rural supplies	460	84.8	75.0	Sandhu et al. (1979)
Colorado rural supplies	164	41.3	—	Ford et al. (1980)
Community water supply study	621	9.0	2.0	Allen and Geldreich (1975)
Tennessee-Georgia rural water supplies	1,257	51.4	27.0	Allen and Geldreich (1975)
Interstate highway drinking-water systems	241	15.4	2.9	Allen and Geldreich (1975)
Umatilla Indian Reservation	498	35.9	9.0	Allen and Geldreich (1975)

mestic wastes (Keswick and Gerba, 1980) and ground water is often the apparent cause of outbreaks of both viral and parasitic illness (Craun, 1986). Contamination of drinking water wells largely occurs in shallow or improperly sealed or placed wells. The degree of this contamination is highly variable, depending on rainfall, the nature of the soil, and other environmental conditions (Gerba and Bitton, 1984).

Pathogenic microorganisms thus clearly exist in most raw water sources in the United States; to protect the public's health, they must be reduced to safe levels. Although waterborne microbial diseases have generally been well controlled and much effort currently centers on chemical contamination, pathogen control remains the major concern for drinking water because of the continued presence of pathogens in raw waters. A recent comprehensive summary of the types of data available on microbiological pathogens in water is presented in Table 1-7.

1.3 Chemical Characteristics of Raw Water

The number of chemical substances that can be found in raw water is virtually unlimited. This section discusses a few chemicals that are representative and significant in terms of health, esthetics, and treatability. First,

Table 1-7. Comprehensive Survey of Data Available on Microbial Pathogens in Water

Organism	Occurrence in Water	Dose Response	Latency	Environmental Persistence	Response to Disinfection	Monitoring Methods	Indicator Pathogen	Environmental Concentration
Salmonella spp.	+	+	+ +	+	+	+	—	+
S. typhi	+	+ +	+ +	+	+	+	+	+
Shigella spp.	+	+	+ +	+	+	+	—	—
V. cholerae	+	+ +	+ +	+	+	+	+	+
E. coli	+	+ +	+ +	+	+	+	—	—
Yersinia spp.	+	—	—	—	—	+	—	—
Campylobacter spp.	+	—	—	—	—	—	—	—
Hepatitis A	—	—	+ +	—	+	+	+	+
Enteroviruses	+	+	+ +	+ +	+ +	+	—	—
Rotavirus	—	—	+ +	—	—	+	—	—
Norwalk Agent	—	—	+ +	—	—	+	—	—
E. histolytica	—	—	+ +	+	+	+	—	—
G. lamblia	+	+	+ +	+	+ +	+	+	+

(—) = few to no data.
(+) = limited data, needs to be augmented.
(++) = adequate data available.
Data adapted from Cooper et al. (1986).

however, this section discusses general properties of raw water that can influence the behavior and exposure implications of chemicals.

The pH of most uncontaminated natural waters is controlled by the carbon dioxide content and other minerals and ranges between 5.5 and 8.5. This is an important consideration, primarily because pH level sometimes determines whether a chemical is a potential health concern.

Alkalinity is a measure of acid-buffering strength. In natural waters, the level is controlled by the soluble carbonate and bicarbonate content. Ground waters tend to be higher in alkalinity than surface waters, an important difference that affects levels of dissolved metals. High alkalinity (greater than 50 mg/L as calcium carbonate), for instance, tends to inhibit corrosion.

Hardness of water indicates the level of soluble calcium and, to a lesser extent, magnesium salts. Ground waters tend to have higher hardness than surface waters because of their high levels of dissolved carbon dioxide and relatively low pH, which causes limestone to dissolve in the aquifer.

Ground waters also tend to be in a more oxygen-deficient (anaerobic) state, whereas surface waters are in a more *aerobic* (oxygen sufficient) state.

Most raw water supplies are relatively *homogeneous;* however, eutrophic reservoirs—i.e., those that are rich in dissolved nutrients and seasonally deficient in oxygen (particularly those in the southern states)—are highly *stratified,* with a layer of aerobic surface water overlying anoxic water. In such reservoirs, sulfur, which is present as sulfate at the surface, may be in the form of hydrogen sulfide at greater depths. Thus, municipal water drawn from the bottom of such a source may have a foul odor and require treatment.

Turbidity, a measure of light transmission through water, is historically a water supply characteristic of primary importance. Turbidity is caused by suspended solids, i.e., soil, sediments, and organics, particularly humic substances and algae. Moving waters (especially alluvial rivers) tend to be most turbid; quiescent reservoirs are less turbid, and naturally filtered ground water generally has insignificant turbidity. Turbid waters can interfere with disinfection and, possibly, contain precursors to toxic trihalomethanes.

1.3.1 Minerals

Minerals are found in all natural waters (see Table 1-8). They are referred to collectively as total dissolved solids (TDS). High TDS levels, i.e., those exceeding 500 mg/L, are encountered in some coastal aquifers (due to seawater intrusion), in some rivers receiving irrigation discharge, and in many wells as a result of natural salt deposits. The TDS typically include calcium, magnesium, sodium, chloride, and sulfate. Two lesser contributors to TDS are silica, usually present at less than 20 mg/L, and potassium, usually present at no more than 10 mg/L.

Table 1-8. Occurrences of Inorganic Contaminants in Excess of the SMCLs[a] in Public Drinking Water Sources—AWWA[b] Survey Data Only

	Chloride		Copper		Iron		Manganese		Sulfate		Total Dissolved Solids		Zinc	
	A[c]	E[d]	A	E	A	E	A	E	A	E	A	E	A	E
SMCL, mg/L	250		1.0		0.3		0.05		250		500		5.0	
Alabama	4					20		14			4			
Delaware	2	0									0			
Georgia	1		0		4				0		0		0	
Hawaii	4				75		75				15			
Idaho		5		5		200		200		10		10		5
Indiana	1				103		81		14					
Kentucky	15		0		24		36		63		14		0	
Maine					25		34						0	
Massachusetts	2		21		152		382							
Michigan	36		6		632	380	290	435	26	130	269	250	2	
Minnesota	5													
Missouri	17		2		130		70		8		101			
Montana	2				36		52		82		185	200		
Nebraska	4					110		75	6					
New Hampshire	0		0						0				0	
New Mexico	23				20		20		196		382			
New York		10		0	155	200	150	50		5		3		0
North Dakota	32	12	6		127		168		127		240		0	0
Ohio				25		515		515		230		35		
Oklahoma	17		0		64		64		13		147		0	
South Dakota	25		0		158		207		211		281			
Texas	295			10	479		360		283		1387			2
Vermont	4				40		45							

Table 1-8. (continued)

	Chloride		Copper		Iron		Manganese		Sulfate		Total Dissolved Solids		Zinc	
SMCL, mg/L	A[c]	E[d]	A	E	A	E	A	E	A	E	A	E	A	E
	250		1.0		0.3		0.05		250		500		5.0	
Washington					181		238					130		
West Virginia		30				250		300		40				
Wisconsin	3		24		314		131		22		177		1	
Total	193	356	59	40	2200	2194	1998	1994	768	698	1815	2015	3	7

[a]SMCL = Secondary maximum contamination level.
[b]AWWA = American Water Works Association (1985). Data were collected from individual states that reported data from their local municipal utilities from June, 1981 through January, 1984. The number of samples was not given in the reports received from the states. Blank spaces indicate that no report was received from the state for this contaminant.
[c]A = actual.
[d]E = estimated.

1.3.2 Heavy Metals

Heavy metals of notable concentrations in public water supplies are arsenic, barium, cadmium, chromium, copper, iron, lead, manganese, mercury, selenium, and zinc. Table 1-9 shows the regional distribution of some of these metals, indicating where contamination is in excess of the Maximum Contaminant Level (MCL)—the regulatory level promulgated by EPA. High levels of cadmium, chromium, lead, and mercury are more commonly associated with anthropogenic (i.e., human) than natural sources.

Cadmium levels in raw waters may be due to storm run-off from municipal and industrial areas. *Chromium* is present in raw water; its chemical form depends on the oxidation-reduction potential of the water. In anoxic waters (typically ground water), chromium is present in the trivalent state, whereas hexavalent chromium predominates in aerobic (generally surface) waters. Trivalent chromium is a nutritionally essential element, but hexavalent chromium causes adverse health effects. *Lead* rarely occurs in raw water at high levels; it is most commonly a problem of leaching from service lines or consumer's plumbing. *Mercury* occurs in natural waters as inorganic mercury salts and as methylmercury.

The following metals may reach substantial levels in raw natural waters: *arsenic, barium, copper, iron, manganese, selenium,* and *zinc.* Iron and manganese are common in raw water. In anoxic ground waters, they are present in the more soluble, reduced forms. When these waters are exposed to air, the metals are oxidized to less soluble forms, which can result in murky drinking water and stained laundry. In excess, copper and zinc give a metallic taste to water; their occurrence usually results from corrosion in the distribution system rather than from natural occurrence in a raw water source.

Selenium is a proven essential nutrient to humans. Arsenic, on the other hand, is a possible trace nutrient to animals. Its essentiality to humans has not been determined. Both can have toxicological consequences depending on their levels. The number of reported instances in which the MCL for each of these elements is exceeded (by state) is reported in Table 1-9. Barium levels have also been reported in Table 1-9 by state in which the MCL was exceeded.

1.3.3 Other Nonorganic Substances

Aluminum is a ubiquitous metal that occurs in natural waters. (Aluminum salts are also used to purify water and, depending on the treatment conditions, some aluminum will remain after treatment.)

Some raw waters (e.g., western Lake Superior and a significant part of the West Coast) contain high levels of *asbestos*. (Asbestos contamination may

Table 1-9. Occurrences of Primary Inorganic Contaminants in Excess of the Maximum Contaminant Levels (MCLs) in Public Drinking Water Sources[a]

MCL (mg/L)	As 0.05	Ba 1.0	Cd 0.010	Cr 0.05	Pb 0.05	Hg 0.002	NO_3 10.0	Se 0.01
Alabama	0	0	0	0	0	0	0	0
Alaska[c]	1	1	1	1	4	2		
Arizona	4	0	1	0	3	0	16	0
Arkansas[c]								
California	3	0	0	0	0	0	8	1
Colorado	3	0	0	0	0	1	12	5
Connecticut[c]					5		2	
Delaware	0	0	0	0	1	0	13	0
Florida[c]					1		2	
Georgia	0	1	0	0	0	0	1	0
Hawaii	0	0	0	0	0	0	0	0
Idaho	0	0	0	0	0	0	3	0
Illinois	3	14	0	0	0	0	11	0
Indiana	0	1	0	0	0	0	1	0
Iowa	1	1	0	0	0	0	39	5
Kansas[c]	0	0	0	0	0	0	45	21
Kentucky	0	0	0	0	0	0	0	0
Louisiana[c]				13				
Maine	0	0	0	0	2	0	0	0
Maryland	0	0	0	0	0	0	6	0
Massachusetts	0	0	0	0	1	0	0	0
Michigan	1	0	0	0	0	0	6	0
Minnesota	1	0	0	0	0	0	7	0
Mississippi[c]								
Missouri	0	1	0	0	0	0	2	0
Montana	3	0	1	0	7	11	10	2
Nebraska	0	0	0	0	0	0	12	19
Nevada	16					4	2	
New Hampshire	4	0	0	0	0	0	0	0
New Jersey								
New Mexico	1	0	0	0	0	0	2	1
New York	0	4	0	0	0	0	12	0
North Carolina	2	0	0	0	1	0	5	2
North Dakota	1	0	0	0	0	0	0	0
Ohio	0	5	0	0	1	0	3	1
Oklahoma	0	2	0	2	0		19	10
Oregon	0	0	0	0	0	0	1	0
Pennsylvania	0	7	5	0	7	5	35	1
Rhode Island					1	1		
South Carolina		10	5	14	15	7	1	2
South Dakota	1	0	0	0	0	0	10	4
Tennessee	0	1	1	1	0	0	0	0
Texas	1	0	0	0	0	0	36	9
Utah		1					2	
Vermont	0	0	0	0	0	0	2	0
Virginia							1	1

Table 1-9. (continued)

MCL (mg/L)	Inorganic Contaminant[b]							
	As 0.05	Ba 1.0	Cd 0.010	Cr 0.05	Pb 0.05	Hg 0.002	NO₃ 10.0	Se 0.01
Washington	0	0	0	0	6	0	20	0
West Virginia		3	0	0	0	0	0	3
Wisconsin	0	0	0	0	0	0	18	0
Wyoming	0	0	0	0	0	0	4	1
TOTAL	46	52	14	18	55	30	369	88

[a]Data from EPA (collected from October, 1977 through September, 1981) and American Water Works Association (collected from June, 1981 through January, 1984). The number of samples was not given in the reports received from the states. Blank spaces indicate that no report was received from the state for this contaminant.

[b]As—arsenic, Ba—barium, Cd—cadmium, Cr—chromium, Pb—lead, Hg—mercury, NO₃—nitrate, Se—selenium.

[c]EPA data only.

also result when corrosive waste distributed by asbestos cement pipe leaches asbestos.) However, it is not a widespread problem in raw water sources.

Fluoride is a naturally occurring halide found in raw water supplies throughout the United States. Calcium fluoride is highly insoluble, so raw waters of normal hardness usually do not contain high levels (greater than or equal to 4 mg fluoride/L). However, some ground waters and surface waters in the Southwest and Southeast have exceptionally high levels of this element.

Nitrate usually occurs in natural waters at levels less than 1 mg/L (calculated as nitrogen). Higher levels in either surface or ground waters may indicate contamination from an anthropogenic (human) source, such as fertilizers, feed lots, or domestic sewage. All nitrate salts are highly soluble in water. There is a natural equilibrium between nitrate and nitrite, controlled by the oxidation-reduction potential, in which nitrate generally predominates.

Radionuclides are of both natural and anthropogenic origin and are found in both surface and ground waters. Radium, radon, and uranium are the ones that occur at significant levels. Radium is the heaviest of the alkaline earth metals, which include magnesium, calcium, and barium. The predominant isotopes are 226[Ra] and 228[Ra]. Radon, a water-soluble gas, is a decay product of radium. Its volatility and short half-life (3.8 days for 222[Ra], the most stable isotope) make it only a minor health concern in surface waters. However, in ground waters where it can be transported to indoor air and inhaled, levels are much higher, with values exceeding 1,000 pCi/L in some northeastern states. Uranium is a member of the actinide family of rare metals and is more commonly found in ground waters than are radon and radium. Its chemical state depends on the oxidation-reduction potential. In aerobic waters, it exists in a highly soluble hexavalent state; in oxygen-deficient waters, it exists in a relatively stable tetravalent form. The predominant isotopes in raw water are 234[U] and 238[U].

1.3.4 Organics

Significant organics in water include both natural and synthetic chemicals. The maximum levels reported for synthetic organics are presented in Table 1-10. The reported numbers should not be interpreted as a general problem throughout the United States but rather as reflecting problems in some community water supplies, e.g., pesticides in agricultural areas. Natural organic materials are found in surface waters throughout the United States at levels of 1 to 20 mg/L, which far exceed levels of synthetic chemicals.

Natural Organics. Natural organics come from soil run-off, forest canopy drip, aquatic biota, decomposition of vegetative biomass, and human and animal wastes. Humic substances are important as sources of turbidity. Fulvic acids, which commonly cause color in raw water, occupy the lower end of the molecular weight scale; this fraction includes most of the so-called *trihalomethane precursors*, i.e., those organics that produce haloforms, principally trichloromethane, when present in water treated with chlorine. Two common algal metabolites, with odors detected at levels below 20 ng/L, are geosmin and methylisoborneol; they occur in eutrophic surface waters and are not removed by conventional treatment.

Synthetic Organics. Synthetic organic materials in water include pesticides and many other manmade compounds, most frequently chlorination by-products.

(a) Pesticides. Pesticides are chemicals applied to crops, forests, soils, homes, gardens, lawns, livestock, pets, and finished products (e.g., lumber) to control the adverse health, esthetic, and economical effects of insects and other arthropods, nematodes, bacteria, fungi, algae, weeds, brush, rodents, and occasionally birds, fish, and larger animals. These substances can contaminate surface waters through run-off, especially from agricultural areas, and ground water by leaching downward from disposal areas and other areas of high concentration. Table 1-10 presents data on pesticides that are currently of concern to EPA.

Alachlor, a slightly soluble herbicide used on corn and soybeans, has been detected in both surface and ground waters at levels in the μg/L range. The carbamate aldicarb (Temik) and carbofuran have been detected in ground waters at levels up to 50 μg/L.

Chlordane is a highly chlorinated hydrocarbon of low water solubility (less than 1 mg/L); it is no longer registered by EPA for manufacture, sale, or distribution, and all use of existing stock must have ceased by April 15,

Table 1-10. Maximum Reported Levels of Some Synthetic Organics in Water Supplies[a]

	Surface Water	Ground Water	MCLG[b]	MCL[c]
	µg/L		(µg/L)	(µg/L)
Acrylamide	no data	no data	0	
Alachlor[f]	104	16	0	
Aldicarb (Temik)[f], including sulfoxide and sulfone	detected	50	10	
Carbofuran[f]	—	50	36	
Chlordane[f]	—	0.2	0	
Chlorobenzene	—	5	60	
Dibromochloropropane[f]	—	60	0	
o-Dichlorobenzene	est. ≥ 0.5[d]		620	
cis/trans 1,2-Dichloroethylene	est. ≥ 20[d]		70	
1,2-Dichloropropane[f]	mean 8.7[d]		0	
2,4-D[f]	1.1	—	70	100
Epichlorohydrin	no data	no data	0	
Ethylbenzene	—	2,300	700	
Ethylene dibromide[f]	—	580	0	
Heptachlor[f]	—	1	0	
Heptachlor epoxide[f]	—	—	0	
Lindane[f]	—	>0.01	0.2	4
Methoxychlor[f]	50	—	340	100
Polychlorinated biphenyls (PCBs)	—	1.27	0	
Pentachlorophenol[f]	12	—	220	
Styrene	none detected		0	
Toluene	2,500[e]		2,000	
Toxaphene[f]	detected	detected	0	5
2,4,5-TP (Silvex)[f]	0.08	0.3	52	10
Xylene	5.2	750	12,000	

[a]EPA (1985).
[b]MCLG = Maximum Contaminant Level Goal (EPA).
[c]MCL = Maximum Contaminant Level. Interim MCLs are given for five chemicals. The Agency plans to propose MCLs for these 25 chemicals (including the 5 that are interim MCLs) by summer 1988. The Agency plans to propose final MCLs for all 25 chemicals by June 1989.
[d]Maximum not available.
[e]Source not identified.
[f]This compound is a pesticide.

1988. The chemical has been detected in well water at the 0.01 µg/L level. Dibromochloropropane (DBCP), a nematocide formerly used on strawberries and other crops, has been detected at the µg/L level in drinking water. The moderately soluble halocarbon 1,2-dichloropropane, used as a fumigant for nematodes and as an industrial solvent, has been detected at the µg/L level in wells. The systemic herbicide 2,4-D is a highly soluble, readily degradable compound used to control terrestrial and aquatic broadleaf weeds; it has been detected in surface water at levels of about 1.0 μg/L.

Although most uses of heptachlor and heptachlor epoxide have been suspended, heptachlor has been detected at μg/L levels in ground-water supplies. Methoxychlor is an insecticide chemically related to DDT. It has been detected in drinking water in high application areas that use surface water as the raw water source.

Pentachlorophenol (Penta) is a slightly soluble herbicide, insecticide, and fungicide used mainly as a wood preservative. Although most of its agricultural uses have been cancelled, it has been detected in surface waters (but not in ground waters).

Other Anthropogenic Organics. These include thousands of synthetic materials besides pesticides in common use. These chemicals may enter water supplies by leaching from disposal areas, precipitation from the atmosphere, or discharge from a sewage or industrial treatment plant. Examples of synthetic chemicals (in addition to pesticides) of concern to EPA are included in Table 1-10.

REFERENCES

Allen, M.J., and E.E. Geldreich. 1975. Bacteriological criteria for ground water quality. *Ground Water* 13:45–51.

AWWA (American Water Works Association). 1985. An AWWA survey of inorganic contaminants in water supplies. *J. Am. Water Works Assoc.* 77(5):67–72.

Centers for Disease Control. 1985. Centers for Disease Control Water-Related Disease Outbreaks Surveillance. *In:* U.S. Dept. of Health and Human Services, Annual Summary, 1984. DHHS Publication no. (CDC) 99–2510. Atlanta, Georgia: Centers for Disease Control.

Cooper, R.C., A.W. Olivieri, R.E. Danielson, P.G. Badger, R.C. Spear, S. Selin. 1986. Evaluation of military field-water quality. *In:* Infectious organisms of military concern associated with consumption: Assessment of health risks and recommendations for establishing related standards, vol. 5, UCRL-53709. Berkeley, California: Lawrence Livermore National Laboratory, p. 2817.

Council on Environmental Quality. 1979. Environmental Quality—1979. Washington, D.C.: Council on Environmental Quality.

Craun, G.F. 1986. Waterborne diseases in the United States. Boca Raton, Florida: CRC Press.

Craun, G.F. 1979. Waterborne disease—a status report emphasizing outbreaks in ground water. *Ground Water* 17:183–191.

EPA. 1986. Bacteriological ambient water quality criteria; availability. *Fed. Regis.* 51:8012. March 7.

EPA. 1985. National primary drinking water regulations: synthetic organic chemicals, inorganic chemicals and microorganisms. *Fed. Regis.* 50:46935–47622. November 13.

EPA. 1978. Guidance for planning the location of water supply intakes downstream from municipal wastewater treatment facilities. Report submitted by Culp/Wesner/Culp, El Dorado Hills, California, to EPA Office of Drinking Water, Washington, D.C.

Feachem, R.G., D.J. Bradley, H. Garelick, D.D. Mara. 1983. Sanitation and disease: Health aspects of excreta and wastewater management. New York, New York: John Wiley & Sons, Inc.

Ford, K.L., J.H.S. Schoff, T.J. Keefe. 1980. Mountain residential development minimum well protective distances—well water quality. *J. Environ. Hlth.* 43:130–133.

Geldreich, E.E. 1986. Potable water: new directions in microbial regulations. *Am. Soc. Microbiol. News* 52(10):530–534.

Gerba, C.P. 1984. Strategies for the control of viruses in drinking water. Report to the American Association for the Advancement of Science, Environmental Science and Engineering Program, Washington, D.C.

Gerba, C.P. and G. Bitton. 1984. Microbial pollutants, their survival and transport pattern to groundwater. *In:* Bitton, G. and C.P. Gerba (eds.), Ground water pollution microbiology. New York, New York: John Wiley & Sons, Inc, pp. 65–88.

Jakubowski, W. 1984. Detection of *Giardia* cysts in drinking water. *In:* Erlandsen, S.L. and E.A. Meyer (eds.), *Giardia* and Giardiosis. New York, New York: Plenum Press, pp. 263–286.

Keswick, B.H. and C.P. Gerba. 1980. Viruses in ground water. *Environ. Sci. Technol.* 14:1290–1297.

Lippy, E.C. and S.C. Waltrip. 1984. Waterborne disease outbreaks—1946–1980: A thirty-five year perspective. *J. American Water Works Assoc.* 76:60–67.

MMWR. 1983. Diarrheal diseases control program. *Morbidity and Mortality Weekly Report* 32:73–75.

Rendtorff, R.C. 1954. The experimental transmission of human intestinal protozoan parasites. II. *Giardia lamblia* cysts given in capsules. *Am. J. Hygiene* 59:204–220.

Rose, J.B., S. Kayed, M.S. Madore, C.P. Gerba, M.J. Arrowood, C.R. Sterling, J.L. Riggs. 1987. Methods for the recovery of *Giardia* and *Cryptosporidium* from environmental waters and their comparative occurrence. Proceedings: Calgary *Giardia* Conference. In press.

Rose, J.B. and C.P. Gerba. 1986. A review of viruses in treated drinking water. *Curr. Pract. Environ. Sci. Eng.* 2:119–140.

Sandhu, S.S., W.J. Warren, P. Nelson. 1979. Magnitude of pollution indicator organisms in rural potable water. *Appl. Environ. Microbiol.* 37:744–749.

Sobsey, M.D. and B. Olson. 1983. Microbial agents of waterborne disease. *In:* Berger, P.S. and Y. Argaman (eds.), Assessment of microbiology and turbidity standards for drinking water. EPA 570/9–83–001. Washington, D.C.: U.S. EPA.

Todd, E.C.B. 1987. Impact of spoilage in foodborne diseases on national and international economies. *Int. J. Food Microbiol.* 4:83–100.

Walsh, J. and K. Warren. 1979. Selective primary health care: An interim strategy for disease control in developing countries. *N. Engl. J. Med.* 301:967.

Ward, R.L. and E.W. Akin. 1984. Minimum infectious dose of animal viruses. *CRC Crit. Rev. Environ. Control* 14:297–310.

White, G.F. and D.J. Bradley. 1972. Drawers of water. Chicago, Illinois: University of Chicago Press.

WHO (World Health Organization). 1984. The role of food safety in health and development. Technical report series no. 705. Report of a Joint FAO/WHO Expert Committee on Food Safety. Geneva, Switzerland: World Health Organization.

WHO (World Health Organization). WHO International Reference Center for Community Water Supply, Annual report. Rijswijk: The Netherlands. World Health Organization.

2

Process Trains

SUMMARY

Water treatment processes provide barriers, or lines of defense, between the consumer and waterborne disease. These lines of defense can be either single, such as when a locality provides disinfection only, or multiple. The most common treatment for surface water supplies—conventional treatment—consists of coagulation, flocculation, sedimentation, filtration, and disinfection. Lime or lime-soda softening processes are widely used in the Midwest and the South to treat some of the harder waters resulting from exposure to limestone formations.

The principal reasons for treatment of surface waters are to reduce particulates, to reduce health risks from microbiological contaminants, and to improve esthetic values such as color and taste. Microbiological contamination of ground waters can occur, but it has historically been regarded as a less important issue than surface water contamination, and treatment technologies for ground water have focused on improving the esthetic quality. Nevertheless, many states do require chlorination of ground water as a means of providing additional microbiological protection.

More communities are beginning to use treatment processes other than disinfection to remove inorganic and organic contaminants. In general, smaller communities face greater challenges than larger ones in meeting water quality standards. The per capita costs of drinking water treatment are much greater for smaller facilities, and this may limit their treatment options. In addition, small systems frequently have less latitude in seeking alternative water supplies and are thus more likely to experience local water quality problems.

2.1 Introduction

The level of drinking water treatment in the United States varies widely depending on such factors as the purity of the water source and the size of population served. At the simplest level, source water is passed on to the consumer untreated. This is a common practice with ground water. The next level of treatment is simple disinfection, usually with chlorine or chloramines. Many utilities with relatively protected surface waters as their source use this level of treatment. This level of treatment can be sufficient if the raw water is of very high quality.

Commonly, however, treatment plants use several different processes to treat drinking water. The combination of processes is referred to as a process train (Figure 2-1). The simplest process train is chlorination followed by filtration through sand or coal. *Filtration* removes particulate matter from the water and reduces turbidity. At the next level of treatment—*in-line filtration*—a coagulant is added prior to filtration. Coagulation alters the physical and chemical state of dissolved and suspended solids and facilitates their removal by filtration. More conservative utilities add a flocculation (stirring) step before filtration which enhances the agglomeration of particles and further improves the removal efficiency. This practice is called *direct filtration*. In direct filtration, disinfection is enhanced by adding chlorine (or an alternative disinfectant) at the end of the process train as well as at the beginning.

By far the most common process train for surface water supplies is *conventional treatment*, which consists of disinfection, coagulation, flocculation, sedimentation (the gravitational settling of heavier particles), and filtration, followed by a second disinfection step. Often, additional steps such as preoxidation, preaeration, adsorption, and/or presedimentation are added. Conventional treatment processes are described in more detail in Chapter 4. Additional information can be found in standard texts such as Montgomery (1985), Sanks (1978), and Hudson (1981).

Lime or lime-soda softening are additional processes that are widely used in the Midwest and the South to treat the hard waters that originate from limestone formations and other sources. In conventional lime softening, sufficient lime is added to convert most of the alkalinity in the water to carbonate ion; this precipitates the calcium. The process generally produces a clearer effluent if the pH is sufficiently increased to ensure that magnesium hydroxide also precipitates. Because high pHs and oxidants (e.g., dissolved oxygen, chlorine, ozone, or permanganate to oxidize iron or manganese) are

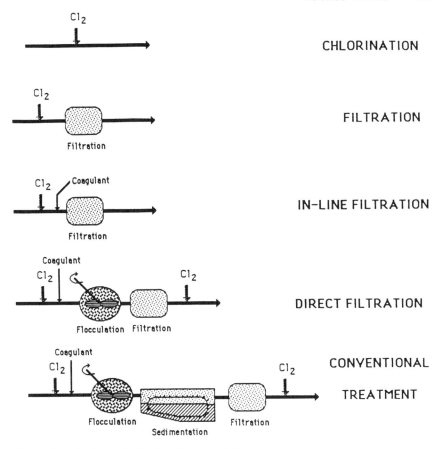

Figure 2-1. Typical water treatment process trains.

often used during softening, the process also provides a certain amount of disinfection.

Several specialized processes have been developed to remove specific contaminants in ground water that affect the suitability of the water for domestic use (e.g., taste, odor, and staining problems). Besides lime softening, common techniques include iron, manganese, and sulfide removal. Both sulfide and iron removal are most often accomplished via aeration and/or chlorination followed by filtration. When manganese is also present, pH adjustment and stronger oxidants or special media are sometimes used.

2.2 Reasons for Raw Water Treatment

Historically, the principal reasons for treating surface waters were to reduce particulates, eliminate the health risk from microbiological contaminants, and improve certain esthetic values such as color and taste. In meeting these primary objectives, many inorganic and organic contaminants that may be present are also removed to a certain degree.

Surface waters have always been more vulnerable to microbial contamination than ground waters, due to such urban-related inputs as sewage and run-off. Ground waters, on the other hand, are more protected from microbial influxes by the mere fact that they are underground. Microbial infestations that do occur are generally related to improperly sealed or placed wells. Nevertheless, many states do require chlorination of ground water as a means of providing additional microbiological protection. Some require disinfection only if monitoring indicates that the source water may have high levels of microorganisms; others require disinfection across the board. According to the American Water Works Association Research Foundation, four of every five large ground-water supplies surveyed were disinfected (AWWARF, 1987).

Though microbiological contamination of ground waters can and does occur (see Table 1-2 in Chapter 1), it has historically been regarded as a less important issue than surface water contamination, so treatment technologies have focussed more on improving the esthetic quality of the water than on disinfection. Iron and manganese are removed because they can stain laundry, and sulfides are removed to prevent unacceptable odors ("rotten egg" smell).

Recently, treatment plants have begun to use additional processes to remove inorganic and organic contaminants. For example, volatile organic solvents such as trichloroethylene and tetrachloroethylene are sometimes (approximately 20 percent) found in ground water at concentrations of about 0.1 to 100 µg/L (sometimes much higher in private wells and hazardous waste sites) and are generally removed by air stripping and/or granular-activated carbon.

2.3 Relative Population Exposures to Treatment Chemicals/Technologies

Table 2-1 gives estimates of the U.S. population that receives water treated by each of the available technologies. These estimates are derived by combining two separate data sets: (1) treatment profiles of community water systems by the size category, i.e., the percent of water systems in each size category using a specific type of treatment (EPA, 1982), and (2) size distribution of community water systems and the population served by each size category (EPA, 1986). Noncommunity water systems (e.g., campground, gas stations, schools, workplace, hospitals, etc.) are not included in these calculations as similar data could not be found on these systems. It is suggested that data in Table 2-1 be used only for relative comparisons of popula-

Table 2-1.   Estimates of Relative Population Exposures to Water Treated by Different Technologies[a]

Treatment Technique	Population Served (Millions)
1. Disinfection	
Ammonia and chlorine (chloramine)	40.1
Chlorine (liquid gas chlorine and hypochlorite)	170.8
Other disinfection	6.2
2. Conventional plant (coagulation, sedimentation, filtration)	107.7
3. Direct filtration	34.6
4. Fluoride addition	95.7
5. Corrosion control	94
6. Activated carbon (granular and powdered)	82.4
7. Aeration	28.5
8. Lime soda softening	25.5
9. Iron removal	33.1
10. Reverse osmosis	0.35
11. Activated alumina	3.7
12. Ion exchange	3.7
13. Other	29.2

[a]Estimates based on 1982 treatment profile information on community water systems and 1985 data on the population served by drinking water systems of various sizes. Estimates derived from data in EPA (1982) and EPA (1986).

tion served by different technologies since these numbers are derived by combining the 1982 treatment profile data with 1985 estimates of the population served by these systems. As might be expected, disinfection is the most common treatment process, followed by coagulation, fluoride addition, corrosion control, and activated carbon treatment. Among the disinfectants, chlorine is the most common (an estimated 171 million people are served chlorinated water), followed by chloramine and other disinfectants. Some of the least commonly used treatment methods are: reverse osmosis, ion exchange, and activated alumina.

2.4 Relative Costs of Drinking Water Treatment Technologies

Table 2-2 summarizes the total treatment costs for several treatment processes, including filtration, disinfection, coagulation, flocculation, and sedimentation (EPA, 1987). The costs are broken down by size of plant, ranging from those that serve 25 to 100 people to those serving over 1 million people. There is a large disparity in cost between small and large facilities. Chlorination, for instance, costs 65.9 cents/1,000 gallons for the smallest facilities and 0.7 cents/1,000 gallons for the largest facilities. This trend holds true for all types of filtration, disinfection, and supplemental processes (i.e., chemical additions). Disinfection processes for small facilities are (in order of increasing treatment cost) ultraviolet light, chlorine, ozonation, chloramination, and chlorine dioxide. For the largest plants the order changes to chlorine, chloramination, chlorine dioxide, and ozonation.

While the difference between the highest and lowest costs is only 5 cents/1,000 gallons for the largest size category, that difference becomes appreciable when translated to over a million gallons/day. Additionally, although the above cost estimates were valid for 1987, they may have changed depending upon technological improvements.

Table 2-3 summarizes ground-water disinfection costs. Chlorine dioxide and chloramination are expensive for small plants; for very large plants all disinfection methods show a relative parity of costs. This parity for large facilities is demonstrated by analyses performed for the City of Los Angeles to evaluate the costs of various treatment options for controlling turbidity, taste, and odor, reducing THMs, and ensuring bacterial and viral disinfection (Stolarik, 1983). The city's final choice was between ozonation and chlorination. Table 2-4 provides the cost figures for the two processes. By integrating ozonation into the filter plant design, the ozone option resulted in a lower capital cost. Operation and maintenance costs were slightly greater for ozone than chlorine, but in the long run, ozone was the cost-saving option. Comparisons with chlorine dioxide demonstrated that chlorine dioxide would cost twice as much as ozone (Stolarik, 1983). Thus Los Angeles adopted ozonation.

The solutions that are appropriate for large communities are not easily transferable to other communities, especially small ones. Small systems often cannot take advantage of economies of scale in treatment techniques, and thus must pay much greater per capita costs. For example, estimates of the costs for controlling volatile organic hydrocarbons in public water supplies indicate that the monthly per household cost increases for such treatment—independent of the technology selected—are highest in small systems by factors from 2- to 6-fold (Table 2-5). In addition, small systems frequently have less latitude in seeking alternative water supplies and thus are more likely to experience local water quality problems than are larger sys-

Table 2-2. Summary of Total Costs for Drinking Water Treatment Processes

	Total Cost of Treatment (¢/1,000 gallons)											
	Size Category[a]											
	1	2	3	4	5	6	7	8	9	10	11	12
Design Flow (mgd):	0.026	0.068	0.166	0.50	2.50	5.85	11.59	22.86	39.68	109.9	404	1,275
Average Flow (mgd):	0.013	0.045	0.133	0.40	1.30	3.25	6.75	11.50	20.00	55.5	205	650
Treatment Processes[b]												
Filtration[b]												
Complete treatment package plants	944.5	277.4	195.1	113.6	72.8	52.4						
Conventional complete treatment					104.1	70.3	58.6	61.9	53.8	39.3	32.0	31.0
Conventional treatment with automatic backwashing filters					87.9	58.3	50.8	57.6	49.4	41.5		
Direct filtration using pressure filters			322.7	137.2	79.1	48.8	39.2	45.8	36.9	28.2		
Direct filtration using gravity filters preceded by flocculation				150.2	90.5	58.4	46.8	50.5	39.8	28.6	23.6	21.3
Direct filtration using gravity filters and contact basins				131.2	80.9	54.7	44.2	48.0	37.5	26.3	21.4	19.1
Direct filtration using diatomaceous earth	672.9	227.2	134.7	66.6	43.1	43.1	36.1	48.1	41.7	35.4		
Slow-sand filtration	377.8	205.1	133.4	54.7	34.3	28.7	25.3					
Package ultrafiltration plants	455.6	226.8	179.2	138.4								
Disinfection[c]												
Chlorine feed facilities[d]	65.9	23.6	16.2	9.7	4.3	2.8	2.1	1.6	1.3	1.0	0.8	0.7
Ozone generation and feed[e]	109	37.2	27.5	12.7	7.0	4.5	3.4	2.6	2.2	1.7	1.4	1.2
Chlorine dioxide[f]	322	87.7	46.1	16.8	7.0	4.2	2.9	2.2	1.7	1.3	1.0	0.9
Chloramination[g]	163	51.1	23.9	14.4	6.1	3.6	2.6	2.1	1.6	1.3	1.0	0.9
Ultraviolet light	43.2	14.1	8.4	5.4								

Table 2-2. (continued)

	Total Cost of Treatment (¢/1,000 gallons)											
	Size Category[a]											
	1	2	3	4	5	6	7	8	9	10	11	12
Design Flow (mgd):	0.026	0.068	0.166	0.50	2.50	5.85	11.59	22.86	39.68	109.9	404	1,275
Average Flow (mgd):	0.013	0.045	0.133	0.40	1.30	3.25	6.75	11.50	20.00	55.5	205	650
Treatment Processes												
Supplemental Processes												
Add polymer feed, 0.3 mg/L	35.3	11.0	8.2	2.9	2.9	1.2	0.7	0.5	0.3	0.2	0.2	0.1
Add polymer feed, 0.5 mg/L	36.4	11.4	8.4	3.0	3.0	1.4	0.8	0.6	0.4	0.3	0.2	0.2
Add alum feed, 10 mg/L	87.1	26.2	16.5	7.5	1.9	1.1	0.9	0.8	0.7	0.6	0.6	0.5
Add sodium hydroxide feed	30.2	10.7	8.4	4.7	2.7	1.9	1.6	1.6	1.5	1.4	1.4	1.4
Add sulfuric acid feed	27.0	9.9	7.6	4.3	1.0	0.6	0.4	0.3	0.3	0.2	0.2	0.2
Capping rapid-sand filters with anthracite coal	0.5	0.3	0.3	0.3	0.3	0.3	0.3	0.3	0.3	0.3		
Converting rapid-sand filters to mixed-media filters	9.4	5.5	3.3	2.1	2.0	1.7	1.6	1.5	1.5	1.5		
Add tube settling modules	2.7	1.6	0.9	0.7	0.6	0.5	0.4	0.4	0.4	0.4	0.4	0.4
Add contact basins to an inline direct filtration plant	16.1	10.9	6.0	3.6	3.7	2.6	1.8	1.6	1.3	1.0	1.0	1.0
Add rapid mix	91.3	30.1	20.3	7.9	4.0	2.5	2.0	2.0	1.9	1.9	1.8	1.8
Add flocculation	45.2	20.1	13.3	7.7	6.2	3.7	2.3	2.0	1.7	1.3	1.2	1.2
Add clarification	95.5	41.7	36.4	18.3	12.7	11.2	10.7	10.1	9.8	9.5	9.2	9.2
Add hydraulic surface wash	80.1	29.0	13.3	5.5	2.4	1.4	1.1	0.9	0.7	0.7	0.7	0.6
Add filter-to-waste facilities	5.5	4.0	2.4	1.3	0.9	0.4	0.3	0.2	0.2	0.2	0.1	0.1
Finished water pumping[h]	70.3	23.7	12.7	7.0	18.4	16.0	14.9	14.3	13.9	13.7	13.7	13.7

Table 2-2. (continued)

Treatment Processes

Alternatives to Treatment[i]

Construct new well, 350 ft	303.0	117.5	59.5	35.0
Bottled water vending machines	795.5	478.1		

Source: EPA (1987).

[a]Population ranges for each category are:

1. 25–100	7. 25,001–50,000
2. 101–500	8. 50,001–75,000
3. 501–1,000	9. 75,001–100,000
4. 1,001–3,300	10. 100,001–500,000
5. 3,301–10,000	11. 500,001–1,000,000
6. 10,001–25,000	12. >1,000,000

[b]Includes cost for chemical addition and individual liquid and solids handling processes required for operation; excluded are raw water pumping, finished water pumping, and disinfection.

[c]Includes costs for all required generation, storage, and feed equipment; excluded are costs for contact basin and detention facilities.

[d]Dose is 5.0 mg/L; includes hypochlorite solution feed for categories 1–3, chlorine feed and cylinder storage for categories 4–10, and chlorine feed and on-site storage for categories 11–12.

[e]Dose is 1.0 mg/L.

[f]Dose is 3.0 mg/L.

[g]Doses are chlorine at 3.0 mg/L and ammonia at 1.0 mg/L.

[h]Facilities include a package high service pumping station for Categories 1–4, and a custom-designed and constructed station for Categories 5–12.

[i]Design flows are equal to system demand, i.e.,

Category	Design Flow (mgd)	Average Flow (mgd)
1	0.07	0.013
2	0.15	0.045
3	0.34	0.133
4	0.84	0.400

Table 2-3.   Summary of Ground-Water Disinfection Costs

	Total Cost of Treatment (¢/1,000 gallons)[a]											
	Size Category[a]											
	1	2	3	4	5	6	7	8	9	10	11	12
Design Flow (mgd):	0.06	0.14	0.31	0.96	3.06	7.52	15.4	25.3	44.2	124	465	1,505
Average Flow (mgd):	0.013	0.045	0.133	0.40	1.30	3.25	6.75	11.5	20.0	55.5	205	650
Disinfection Method												
Chlorine[b]	138.3	50.8	22.5	10.9	6.2	4.5	3.6	2.9	2.5	1.9	1.4	1.1
Ozone[c]	165.1	61.6	31.6	15.5	7.5	5.0	4.0	3.5	3.1	2.1	1.4	1.2
Chlorine Dioxide[d]	341.0	106.9	50.0	20.1	9.7	5.5	3.7	2.9	2.5	1.7	1.1	0.8
Chloramination[e]	188.4	69.7	34.2	15.7	8.0	5.2	3.7	3.0	2.5	1.9	1.3	1.0
Ultraviolet Light	61.9	20.2	10.1	7.6								
Chlorine[f]	67.6	24.1	13.4	5.5	3.1	1.7	1.1	0.9	0.8	0.6	0.4	0.3
Chlorine Dioxide[f]	320.7	109.6	46.2	16.8	6.3	3.3	2.2	1.7	1.5	0.6	0.6	0.4
Chloramination[g]	158.2	56.5	39.6	9.5	4.1	2.0	1.3	0.99	0.8	0.6	0.3	0.3

[a] See Table 2-1 for definition of size categories.
[b] 2.0 mg/L with 30-minute detention.
[c] 1.0 mg/L with 5-minute contact time (in-pipe).
[d] 2.0 mg/L with 15-minute detention.
[e] 1.5 mg/L chlorine and 0.5 mg/L ammonia with 30-minute detention.
[f] 2.0 mg/L.
[g] 1.5 mg/L chlorine and 0.5 mg/L ammonia.
Source: EPA (1987).

Table 2-4.   Costs of Ozone vs. Chlorine Systems for the City of Los Angeles[a]

Capital Cost Item	Ozone	Chlorine
Process Equipment	5,661,000	166,000
Related Process Structures and Appurtenances	5,461,700	—
Additional Filtering Capacity (related costs)	—	13,070,330
Related Electrical	1,160,000	782,000
Additional Backwash Capacity	—	1,260,500
Miscellaneous	1,094,300	40,000
Subtotal	$13,377,000	$15,318,830
[Construction provision for future facilities (O_3) should C_{12} not achieve taste and odor control, THM standard]		[3,214,000]
Total (including sales tax)	$13,377,000	$18,532,830

Operation and Maintenance Item	Ozone	Chlorine
Chlorine	—	162,400
Additional Chemical Coagulant	—	314,000
Ozone Dielectric Maintenance	127,000	—
Additional Power	498,000	42,100
Total	$ 625,000	$ 518,500

[a]Only items for which there was a cost difference are included. Costs for features that were the same for both processes are not included.
Source: Rice (1985) adapted from Stolarik (1983).

tems. Some treatment techniques are just not financially available to small systems with limited resources. In summary, reasonable alternatives are needed to improve the quality of drinking water for small systems. One possible option is regionalization, wherein several small systems in close proximity share the cost for upgraded treatment trains.

REFERENCES

AWWARF (American Water Works Association Research Foundation). 1987. National trihalomethane survey report. Prepared by Decision Research under the supervision of the Metropolitan Water District of Southern California.

Table 2-5. **Summary of Increased Per-Household Costs for Reducing Trichloro-ethylene (a Volatile Organic Hydrocarbon) Levels in Drinking Water**

Technology	Population Served		
	100–499	1,000–2,499	10,000–24,999
A. 90% Removal (500 mg/L to 50 mg/L)			
Aeration (Packed Tower)	$ 4.65–6.49	$ 2.04–2.58	$ 0.72–0.92
Aeration (Diffused Air)	9.76	5.04	2.41
Adsorption (GAC)	13.48	7.62	2.03
B. 99% Removal (500 mg/L to 5 mg/L)			
Aeration (Packed Tower)	$ 5.11–7.43	$ 2.34–12.00	$ 0.85–1.13
Aeration (Diffused Air)	13.48	8.23	3.81
Adsorption (GAC)	13.95	7.87	6.32

Source: Hacker, 1985.

EPA. 1987. Draft executive summary of technologies and costs for the removal of microbial contaminants from potable water supplies. Washington, D.C.: U.S. EPA Office of Drinking Water, Science and Technology Branch.

EPA. 1986. The national public water system program. FY 85 status report. Washington, D.C.: EPA Office of Drinking Water. May 1986.

EPA. 1982. Survey of operating and financial characteristics of community water systems. Report prepared by Temple, Barker and Sloane, Inc. for EPA. Washington, D.C.: EPA Office of Drinking Water. October 7, 1982.

Hacker, T.L. 1985. Regulatory flexibility and consumer options under the Safe Drinking Water Act. *In:* Rice, R.G. (ed.), Safe Drinking Water. Chelsea, Michigan: Lewis Publishers, Inc., pp. 209–221.

Hudson, H.E., Jr. 1981. Water clarification processes: Practical design and evaluation. New York, New York: Van Nostrand Reinhold.

Montgomery, J.M. 1985. Water Treatment Principles and Design. New York, New York: Wiley Interscience.

Rice, R.G. 1985. Ozone for drinking water treatment—evolution and present status. *In:* Rice, R.G. (ed.), Safe Drinking Water. Chelsea, Michigan: Lewis Publishers, Inc., pp. 123–159.

Sanks, R.L. (ed.). 1978. Water treatment plant design. Ann Arbor, Michigan: Ann Arbor Science.

Stolarik, G. 1983. Ozonation—direct filtration of Los Angeles drinking water. Presented at the International Ozone Association, 6th Ozone World Congress, Washington, D.C.

3

Disinfection

SUMMARY

Disinfection is the most important water treatment process for preventing the spread of infectious disease. For the past several decades, chlorine has been the disinfectant of choice. The discovery and subsequent regulation of some chemical by-products of chlorination (e.g., trihalomethanes) has increased the popularity of other disinfectants such as ozone, chlorine dioxide, and chloramines. Properly applied disinfection provides a critical and effective barrier against waterborne disease. If a water system that used a contaminated source and depended entirely on disinfection were to abandon this treatment, all members in the community would contract one or more waterborne diseases.

Disinfectants can be added either as a first step in the treatment train or after all interfering particles have been removed. The latter is often thought to be the more reliable point for primary disinfection. In addition, disinfectants are often added after treatment to maintain a disinfectant residual in the distribution system. This serves the function of preventing microbial regrowth in the distribution system.

The risks from pathogens due to lack of disinfection are based on numerous human epidemiology data and are quite certain. In contrast, the risks from disinfected water are not as well defined due to a lack of data on chlorination by-products. Existing epidemiologic data on chlorinated water indicate a small cancer risk. Where data from animal studies exist, the quantified risks may be overestimated due to the conservative approach used to assess risk (i.e., use of high-dose animal studies extrapolated to humans using 95 percent upper confidence limits for carcinogens and uncertainty factors of generally 100 to 1,000 for noncarcinogens). However, the possible interactions at low doses cannot be evaluated.

The health risks associated with the use of different disinfectants vary widely, and result from the presence of the disinfectant itself and its by-products in drinking water. When the by-products are considered, questions of safety become much more complex. The amount and nature of these by-products vary greatly depending on the water characteristics.

Lack of data and a lack of biologically valid methodologies for extrapolation from high to low doses and from animal to man are the major problems in estimating risk. More data are needed on the occurrence of pathogens and disinfectant by-products in drinking water, and on the health effects of the by-products of all disinfectants.

Efforts to minimize the health risks associated with chlorination by-products have focused on using alternative disinfectants other than chlorine. Other approaches—such as removing precursors to organic by-products, modifying water conditions, controlling the amount of disinfectant added, and removing the by-products—are also being examined.

3.1 Introduction

Disinfection is the most important process used in water treatment for preventing the spread of infectious diseases. Generally, disinfection is accomplished through the use of an oxidant. For the past several decades, chlorine has been the disinfectant of choice, though some utilities have used chloramines and a limited number have used ozone and chlorine dioxide. The discovery and subsequent regulation of some chemical by-products of chlorination have increased the use of other disinfectants such as ozone, chlorine dioxide, and particularly chloramines.

This chapter assesses the overall health risk associated with disinfection. This assessment considers:

• The reduction of risks from infectious agents.
• The increased toxicological risk due to the disinfectant itself if there is a residual level at the consumer's tap.
• The increased toxicological risks resulting from chemical by-products that are formed.

The rest of this chapter is divided into five sections. The first (3.2) describes the approach to risk assessment. Section 3.3 describes disinfectants that are currently used in public water supplies. Section 3.4 provides information on some of the resulting by-products. Section 3.5 discusses the effectiveness of disinfection in reducing the risk of waterborne infectious disease; Section 3.6 examines the health risks associated with disinfection; and Section 3.7 compares waterborne infection hazards to toxicological hazards that result from the use of alternative forms of disinfection.

3.2 Approach to Risk Assessment

The net change in risk produced by the use of a particular disinfectant is a trade off between reducing the risk of infectious disease and increasing the toxicological risk induced by chemicals. Where appropriate information exists, a quantitative approach has been used to evaluate these risks. These estimates should be considered tentative because they are far from being inclusive of the total microbiological and toxicological risks. Furthermore, the assumptions that underlie risk assessment models used (in this case the linearized multistage model) are difficult to verify experimentally.

The risks from pathogens due to lack of disinfection are based on numerous human epidemiology data and are quite certain. The risks from disinfected water are not as well defined due to lack of data on disinfection by-products. However, existing epidemiologic data (see Section 3.6.1) indicate a small cancer risk from disinfection. Where data exist, the quantified risks may be overestimated due to the conservative approach used to assess risk (i.e., use of high-dose animal studies extrapolated to humans using 95 percent upper confidence limits for carcinogens and uncertainty factors of generally 100 to 1,000 for noncarcinogens).

Risks for carcinogenic effects and for noncarcinogenic effects are generally assessed in very different ways due to differences in assumptions of toxicity. With carcinogens, it is assumed that there is no safe level (threshold) of exposure, i.e., exposure to any amount of a carcinogenic substance could theoretically increase the risk of cancer. With noncarcinogens, it is assumed that there is an exposure threshold below which the chemical causes no adverse health effects. For carcinogens, the risk associated with consumption of 2 liters of water per day over a lifetime is calculated. For noncarcinogens, the Reference Dose (RfD) (formerly called the Acceptable Daily Intake [ADI]) is calculated. The RfD is an estimate (with an uncertainty spanning perhaps an order of magnitude) of a daily exposure to the human population (including sensitive subgroups) that is likely to be without appreciable risk of deleterious effects during a lifetime.

For substances with carcinogenic potential, chemical concentrations are correlated with carcinogenic risk estimates by employing a cancer potency (unit risk) value together with assumptions for lifetime exposure and the ingestion of water. The cancer unit risk is generally derived from a linearized multistage mathematical model with 95 percent upper confidence limits that provides a low-dose estimate of cancer risk. The cancer risk is characterized as a theoretical upper bound estimate. The true risk to humans is not likely to exceed this upper bound estimate and may, in fact, be lower.

While other models may be used for estimating risk (e.g., one-hit, Weibull, Logit, or Probit), the range of risks described by these models has little biological significance unless data are available to support the use of one model over another. EPA, therefore, recommends the use of the linearized multistage model for consistency in approach and to provide an upper bound estimate of the potential for carcinogenic risk.

Some of the risk assessments performed here for carcinogens (except where noted below) are based on an increased incidence of liver tumors in B6C3F1 mice—a species that has a high spontaneous incidence of tumors. Using such data as a clear indication of carcinogenic risk in humans— without any supporting data from another species—is controversial. Only two of the disinfection by-products—chloroform and dichlorobromomethane—have to date been determined to be carcinogenic in more than one species. A potential disinfection by-product, 2,4,6-trichlorophenol, has also been determined to be carcinogenic.

3.3 Disinfectants Now in Use

Chlorine, chloramines, chlorine dioxide, and ozone are used for more than one purpose and at more than one location in a water treatment plant. The most common locations for disinfectant addition are at the inlet or headworks of the water treatment plant (before any other chemicals are added) and after filtration. These two practices, known as preoxidation and post-disinfection respectively, are described in Sections 3.3.1 and 3.3.2.

Although no comprehensive surveys exist that summarize all aspects of disinfection, the recent American Water Works Association Research Foundation (AWWARF) survey (McGuire and Meadows, 1987) summarizes a great deal of data on the frequency of use of the various disinfectants and on doses typically used. The survey included 907 utilities, of which 484 served more than 50,000 customers. Of these large utilities, approximately 263 exclusively used a lake, a river, or a ground-water source as a water supply. In terms of disinfectant choice, 216 used chlorine, 67 chloramines, 16 chlorine dioxide, and 1 ozone. Estimates of the population exposed to various disinfectants are shown in Table 2-1. These estimates show that 79 percent of the U.S. population is exposed to chlorine, 18 percent to chloramine, and 3 percent to other disinfectants. The use of alternative disinfectants continues to become more common because of THMs formed as treatment by-products of chlorination. The amount of each disinfectant used by these utilities is summarized in Table 3-1.

Table 3-1. Chlorine, Chloramine, and Chlorine Dioxide Dosages for Various Drinking Water Sources

Dose mg/L	Lakes		Rivers		Ground Water		Total	
	No.	%	No.	%	No.	%	No.	%
Chlorine								
None	15	15.3	14	18.7	18	20.0	47	17.9
0.1–0.5	3	3.1	1	1.6	17	18.9	21	8.0
0.6–1.0	10	10.2	7	11.5	18	20.0	35	13.3
1.1–1.5	14	14.3	7	11.5	6	6.7	27	10.3
1.6–2.0	12	12.2	12	19.7	13	14.4	37	14.1
2.1–2.5	10	10.2	5	8.2	4	4.4	19	7.2
2.6–3.0	5	5.1	8	13.1	1	1.1	14	5.3
3.1–3.5	8	8.2	1	1.6	1	1.1	10	3.8
3.6–4.0	6	6.1	7	11.5	0	0.0	13	4.9
4.1–5.0	7	7.1	5	8.2	2	2.2	14	5.3
5.1–10.0	7	7.1	5	8.2	4	4.4	16	6.1
>10	1	1.0	3	4.9	6	6.7	10	3.8
Total	98		75		90		263	
Chloramine								
None	70	93.3	48	49.5	77	85.6	195	74.1
0.1–0.5	2	2.7	1	1.0	2	2.2	5	1.9
0.6–1.0	6	8.0	5	5.2	0	0.0	11	4.2
1.1–1.5	6	8.0	2	2.1	2	2.2	10	3.8
1.6–2.0	5	6.7	5	5.2	1	1.1	11	4.2
2.1–2.5	3	4.0	2	2.1	1	1.1	6	2.3
2.6–3.0	2	2.7	5	5.2	2	2.2	9	3.4
3.1–3.5	0	0.0	1	1.0	0	0.0	1	0.4
3.6–4.0	1	1.3	3	3.1	1	1.1	5	1.9
4.1–5.0	1	1.3	2	2.1	4	4.4	7	2.7
5.1–10.0	1	1.3	1	1.0	0	0.0	2	0.8
>10	1	1.3	0	0.0	0	0.0	1	0.4
Total	97		75		90		262	
Chlorine Dioxide								
None	90	93.8	65	86.7	90	100.0	245	93.9
0.1	0	0.0	2	2.7	0	0.0	2	0.8
0.2	0	0.0	1	1.3	0	0.0	1	0.4
0.3	1	1.0	1	1.3	0	0.0	2	0.8
0.7	1	1.0	0	0.0	0	0.0	1	0.4
0.8	0	0.0	2	2.7	0	0.0	2	0.8
1.0	2	2.1	3	4.0	0	0.0	5	1.9
1.3	0	0.0	1	1.3	0	0.0	1	0.4
1.6	1	1.0	0	0.0	0	0.0	1	0.4
3.0	1	1.0	0	0.0	0	0.0	1	0.4
Total	96		75		90		261	

Source: McGuire and Meadows (1987).

3.3.1 Preoxidation

Preoxidation refers to the addition of an oxidant, such as chlorine, at the entrance to a treatment plant. Preoxidation is generally employed to disinfect, to remove odorous compounds and sulfide, to reduce coagulant demand, to oxidize iron and manganese, and to prevent the formation of biological slimes and algae in the downstream treatment processes. Until the mid-1970s, preoxidation was accomplished almost exclusively with chlorine—a process known as prechlorination. In recent years, other oxidants such as chloramines, chlorine dioxide, and ozone have also been used.

Typical doses used for preoxidation depend on the specific chemistry of the water. For any oxidant, the levels of dissolved organic matter, iron, manganese, and sulfides affect the required dose. For chlorination, the level of ammonia is also very important. Chloramines are much less effective than chlorine at removing tastes and odors and at oxidizing other reduced substances (i.e., substances that chemically react with an oxidizing agent), though they do provide limited help in controlling downstream growth. As a result, chloramines are less frequently used as preoxidants. Oxidant dose ranges are given in Table 3-2.

Table 3-2.   Total Oxidant Doses in Preoxidation

Oxidant	Dose Range (mg/L)	Typical Dose (mg/L)
Chlorine	0 to 25[a]	0.5 to 4.0
Chlorine dioxide	0 to 1[b]	0.3 to 1
Ozone	0 to 15[c]	0.3 to 2

[a]The higher prechlorine doses are usually used to oxidize ammonia or high levels of natural aquatic humus.

[b]Chlorine dioxide addition is limited to 1 mg/L per EPA's recommendation in order to prevent excessive levels of chlorate and chlorite.

[c]The higher levels of preozonation are used to oxidize natural aquatic humus in order to remove the color it imparts to the water.

3.3.2 Post-Disinfection

Post-disinfection is usually considered the primary disinfection step because disinfection is thought to happen most reliably after all interfering particles have been removed by filtration, settling, and/or coagulation. Public health officials and water treatment practitioners rely on post-disinfection to ensure that disinfection is satisfactorily accomplished; additional disinfection at other points in the process is also considered beneficial.

In post-disinfection, the disinfectant is typically added immediately following filtration; however, chlorine and chloramines are sometimes added just before filtration to prevent slime growth in the filter media. Post-ozonation is almost always performed after filtration.

In addition to primary disinfection, a secondary disinfectant is often added at the end of the treatment process to maintain an oxidant residual in the distribution system. This practice is known as *residual maintenance*. Residuals of disinfectants prevent slime formation and the subsequent degradation of water quality in distribution piping.

Generally, the doses of disinfectant for post-disinfection are much smaller than those for preoxidation because oxidant demand has been reduced by upstream processes. Typical oxidant doses and residuals are as shown in Table 3-3.

Table 3-3. Post-Oxidant Doses and Residuals

Oxidant	Typical Dose (mg/L)	Residual (mg/L)
Chlorine	0.5 to 3	0 to 1.5
Chloramines	0.5 to 4	0.5 to 4
Chlorine dioxide	0.2 to 1	0.1 to 0.5
Ozone	0.2 to 1.5	none

3.4 Disinfection By-Products

Substantial information is available on the by-products of chlorination, yet better information is still needed. Information on the other three disinfectants is more limited. A brief review of the by-products of each disinfectant follows.

3.4.1 Chlorine

Figure 3-1 shows chlorine doses applied to drinking water based on a recent survey by the AWWARF (1987). The principal by-products of concern are chlorinated organic chemicals that result from reactions between aqueous chlorine and the natural aquatic humic material in the water—material left there from the decay of vegetation in the watershed. The level of natural aquatic humus is generally measured with the *total organic carbon* (TOC) test (APHA, 1985).

Figure 3-2, summarizing data from the AWWARF survey (McGuire and Meadows, 1987), illustrates the distribution of TOC in U.S. water supplies

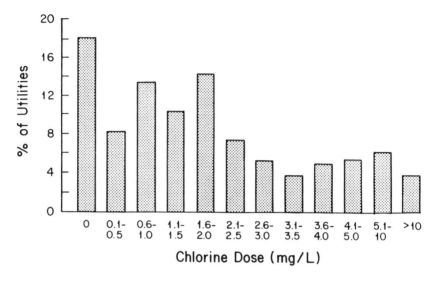

Figure 3-1. Chlorine dose applied to drinking waters. *Source:* Based on data from 139 utilities (AWWARF, 1987).

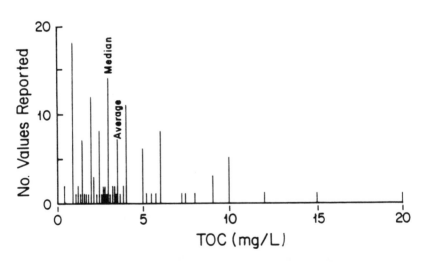

Figure 3-2. Summary of data on total organic carbon in 139 treated waters in the United States. *Source:* Data from AWWARF, 1987.

that reported TOC data. Only 139 of 910 survey respondents provided TOC data, and these were mostly in the greater-than-50,000-population-served category. Of the 139 respondents, 13 reported a TOC of zero, 1 a TOC of 0.1 mg/L, and 5 reported TOCs above 25 mg/L. The remaining respondents fell mainly between 0.1 and 5.0 mg/L TOC. If these are considered outliers, the median is 3 mg/L and the mean is 3.8 mg/L. If the outliers are not discarded, the median remains unchanged, but the mean rises to 6.4 mg/L.

The *total organic halide* (TOX) test (APHA, 1985) is often used as an overall measure of the level of chlorinated organic materials resulting from chlorination; however, relatively few TOX survey data are available and it is not possible to ascribe a specific health risk to any particular level of this surrogate parameter.

The most widely recognized chlorination by-products are the *trihalomethanes* (THMs). In EPA's regulations, the term *total trihalomethane* (TTHM) refers to the sum of chloroform, bromodichloromethane, dibromochloromethane, and bromoform; these are the most common THMs found in chlorinated water. Five major surveys have assessed the THM levels in treated domestic water in the United States.

The AWWARF study covers by far the greatest number of water supplies, but, unfortunately, the data collected in this survey do not include specific information on each THM. This information is necessary to characterize toxicological risk. Probably the most useful surveys for this purpose are the National Organics Reconnaissance Survey (NORS) and the National Organics Monitoring Survey (NOMS-3D). These surveys were taken before the federal regulation on THMs took effect; the AWWARF survey was taken from 1984 to 1986. The median values (about 40 ppb) are similar, but mean values are substantially lower in the more recent (AWWARF) survey. Indeed, Figure 3-3, taken from a summary of that survey (McGuire and Meadows, 1987), shows that most of the differences in concentration between surveys were in systems that reported very high THM levels. The values of the individual THMs obtained in NORS and NOMS are shown in Table 3-4.

Since the 1970s, additional work has been performed to identify the levels of other by-products of chlorination in drinking water supplies. In 1986, EPA surveyed 26 water systems where high levels of THMs were expected (Reding et al., 1986). Although the values obtained are probably higher than the national average, they do offer some perspective on the relative concentration of other possible disinfection by-products. Table 3-5 summarizes the findings of this study.

Other useful data were recently collected by EPA in a survey of 10 cities selected to represent a wide geographic area and several source types (Stevens et al., 1987). Seven compounds were found in every water supply analyzed (see Table 3-6). Six other compounds were found with appreciable frequency (Table 3-7). 1,1,1-Trichloroethane (TCA), trichloroethylene

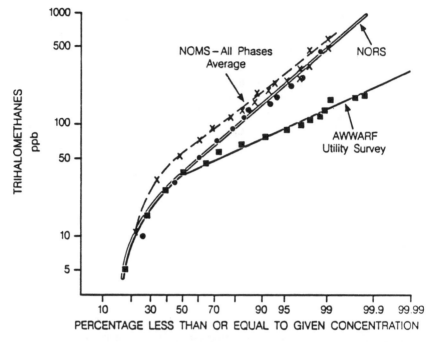

Figure 3-3. Frequency distributions of national THM survey data. During NORS, samples were stored at 2 to 8°C for 1 to 2 weeks prior to analysis. During NOMS-1[a] and NOMS-3T[a], samples were stored at 20 to 25°C for 3 to 6 weeks prior to analysis. During NOMS-3D[a], samples were neutralized with thiosulfate as they were collected. Samples for AWWARF were collected and analyzed per EPA regulations. Latter survey is based upon 12 quarterly averages from 1984 to 1986.
[a]1 = iced; 3T = terminal; 3D = quenched.
Source: McGuire and Meadows, 1987.

Table 3-4.   Summary of NORS and NOMS-3D Trihalomethane Data

	Mean Values (μg/L)	
Compound	NORS	NOMS-3D
Chloroform	21	22
Bromodichloromethane	6	6
Chlorodibromomethane	1.2	2
Bromoform	5	LD[a]

[a]LD = less than detection limit.

Table 3-5. Disinfection By-Products in EPA'S 26-City Survey[a]

Compound	Median Value (μg/L)	Range (μg/L)	No. of Samples Detected/ Analyzed
Chloroform	65.0	<0.7–360	24/25
Bromodichloromethane	8.7	<0.2–77	23/25
Chlorodibromomethane	2.4	<0.2–65	19/25
Dichloracetonitrile	1.0	<0.2–24	21/26
Bromochloroacetonitrile	0.5	<0.2–10	20/26
Chloropicrin	0.3	0.2–1.5	13/26
Bromoform	<0.5	<0.5–54	12/25
Dibromoacetonitrile	<0.3	<0.2–1.8	10/26
Total organic carbon (TOC)	3.6[b]	1.8–8.2[b]	18/18
Total organic halide (TOX)	215.0	10.0–560	18/18

[a]Reding et al. (1986).
[b]TOC is in mg/L.

Table 3-6. By-Products Found in All Samples of EPA'S 10-City Survey[a]

Compound	Median Value (μg/L)	Range (μg/L)	No. of Samples Detected/ Analyzed
Chloroform	28	2.6–594	10/10
Dichloroacetic acid (DCAA)	10–100	from <10->100	10/10
Trichloroacetaldehyde	10–100	from >10-<100	10/10
Chlorodibromomethane	7.6	0.3–31	10/10
Bromodichloromethane	6.8	4.4	10/10
Dichloracetonitrile (DCAN)	2.2	0.2–9.5	10/10
1,1,1-trichloropropanone	<10	All <10	9/9
Total Organic Carbon (TOC)	2.5[b]	<1-10[b]	7/7
Total Organic Halide (TOX)	130	30–1,600	7/7

[a]Stevens et al. (1987).
[b]TOC is in mg/L.

(TCE), tetrachloroethylene (PCE), 3,3-dichloro-2-butanone, and cyanogen chloride were also found at very low levels in some samples. TCA, TCE, and PCE are probably not disinfection by-products, but were instead present in the raw water or leached from water conduit surfaces.

The concentrations of chlorination by-products change as the pH changes, as shown in Figure 3-4 (Croue et al., 1987). Furthermore, the concentrations

Table 3-7.   Other By-Products Found in EPA'S 10-City Survey[a]

Compound	Median Value (μg/L)	Range (μg/L)	No. of Samples Detected/ Analyzed
Chloropicrin	0.4	<0.2–5.6	8/10
Trichloroacetic Acid	<10	ND to 10-100[b]	6/10
Chloroacetic Acid	<10	ND to 10-100[b]	6/10
Bromoform	0.2	<0.2–2.7	6/10
Bromochloroacetonitrile	0.5	<0.2–1.7	6/7
Dibromoacetonitrile	<0.2	<0.2–1.2	3/7

[a]Stevens et al. (1987).
[b]ND = not detectable.

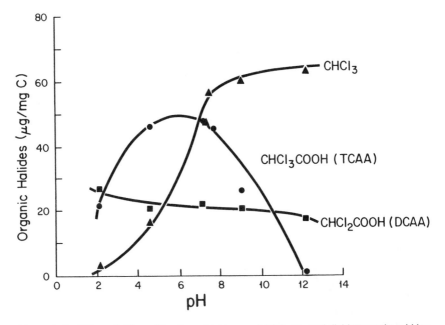

Figure 3-4.   Effect of pH on chloroform, trichloroacetaldehyde, and dichloroacetic acid levels. *Source:* After Croue et al., 1987. Reprinted with permission from "Environmental Science and Technology," Copyright 1987, American Chemical Society.

of chloroform ($CHCl_3$), trichloroacetic acid (Cl_3COOH), and dichloroacetic acid ($CHCl_2COOH$) change with respect to each other. Thus a constant concentration relationship between chlorination by-products cannot be assumed.

Some of the ranges and detected/analyzed values in the above tables differ from those shown in the primary reference (Stevens et al., 1987). The values given in the tables are based on a review of the raw data presented in a letter to participating utilities (Greenberg, 1986). Median values given in the tables for both the 10-city and 26-city survey have been corrected to reflect the entire population surveyed rather than only those where the compounds were found.

Additional perspective can be gained from a survey conducted on North Carolina supplies in 1985 (Norwood et al., 1985). This survey found a median value of 25 μg/L for trichloroacetic acid (TCAA) and 325 μg/L for TOX in 10 water supplies in North Carolina. TCAA accounted for between 1.8 percent and 10 percent of the TOX measured. These latter values are probably higher than the norm; surface waters in North Carolina are typically rather high in TOC.

3.4.2 Chloramines

An early inventory of municipal water supplies (U.S. Public Health Service, 1963) indicated that only 2.6 percent of U.S. supplies surveyed were using chloramines. However, recently many utilities have begun to use chloramines instead of chlorine to avert the formation of THMs. The AWWARF survey (McGuire and Meadows, 1987) found that chloramines were used by 13 percent of smaller lake systems and 25 percent of large surface water systems. The AWWARF survey found that typical doses of chloramines were 1.5 mg/L for systems using lake waters and 2.7 mg/L for systems using surface water from flowing streams. A 1984 survey of some U.S. utilities using chloramines (Trussell and Kreft, 1984) showed a possible distribution of chloramine residuals (see Figure 3-5).

As the chlorine-to-ammonia ratio increases, the two chemicals react first to form monochloramine, then dichloramine, and finally a complex group of by-products, including nitrogen trichloride, nitrogen gas, nitrate, and so on. In most instances the ratio of chlorine to ammonia used in drinking water treatment is low enough to ensure that most of the chlorine is present as monochloramine. The Trussell and Kreft (1984) survey showed that 95 percent of the utilities using chloramines operated at a chlorine:ammonia ratio of 4:1 or less.

The potential for formation of organic by-products during chloramination has been studied less than it has with free chlorine or chlorine dioxide. Chloramines are weaker oxidizing agents than either chlorine or chlorine

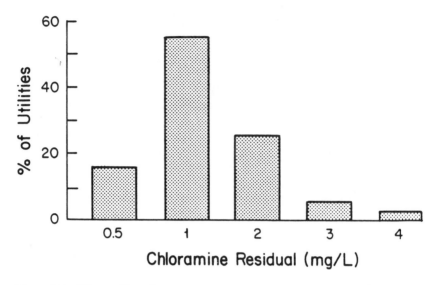

Figure 3-5. Effluent chloramine concentrations. *Source:* Trussell and Kreft, 1984. Based on 27 utilities surveyed. Reprinted from *Chloramination for THM Control: Principles and Practices,* by permission. Copyright 1984, American Water Works Association.

dioxide; however, chloramines hydrolyze to form trace levels of free chlorine, which would react to produce low levels of chlorination by-products. A 1980 report by the National Academy of Sciences (NAS) Safe Drinking Water Committee summarized the data for a large number of compounds that are formed as a result of chlorine substitution, oxidation, amination, and free radical reactions; however, few of these compounds have been looked for in drinking water. The organic reaction by-products that have been isolated in connection with chloramines have largely been the same as those found with free chlorine, but at much lower levels.

How chloramines are added can have considerable impact on the level of organic by-product formation. The most common method used is the separate addition of chlorine and ammonia. Less by-product formation is expected when ammonia is added first (pre-ammoniation); more by-product formation is expected when ammonia is added second (post-ammoniation). Even lower levels of by-product formation can probably be expected when preformed chloramines are used; however, this is not practiced at present. The principal issue of concern when chloramines are used for residual maintenance is probably the chloramine residual itself.

3.4.3 Chlorine Dioxide

As with chloramines, relatively little work has been done to identify the specific by-products formed via reactions with chlorine dioxide. The types of compounds to be expected are acids, epoxides, quinones, aldehydes, disulfides, and sulfonic acids. However, no information on their levels in drinking water exists in the literature.

THMs are not formed when chlorine dioxide is employed as a disinfectant. In a study of the organic reaction products of chlorine dioxide and natural aquatic fulvic acids, only limited reactions were found to occur between chlorine dioxide and aquatic fulvic acid (Colclough et al., 1985). Although not quantified, most of the compounds identified were methyl esters.

In the past, chlorine dioxide generally contained residual chlorine. Thus, there was concern that use of chlorine dioxide would result in the same by-products as free chlorine. Recently, however, processes have come on the market that can produce high-purity chlorine dioxide at reasonable cost. An additional health concern relates to chlorine dioxide's inorganic by-products: chlorate and chlorite ions. Although chlorite is probably of greater health concern, chlorate is difficult to measure; consequently, the health risks of chlorate are difficult to assess. Tests conducted by the EPA suggest that, when chlorine dioxide is added to water, about 50 percent is converted to chlorite ion, 25 percent to chlorate ion, and 25 percent to chloride ion (Symons et al., 1981).

3.4.4 Ozone

Ozone is the most reactive of the oxidants used in water treatment. Ozone reacts rapidly with natural aquatic humus, but relatively little specific information exists about the resulting by-products. The NAS (1980) report lists a number of possible by-products, with particular concern about the possible formation of epoxides; however, these are unstable in water and none has ever been identified. The most likely organic by-products are aldehydes, ketones, and acids. When bromide ion or chlorine is present, ozonation may form some bromate or chlorate.

When humic molecules are ozonated, their structure is probably not destroyed, but a part of their functional groups is oxidized. Most of the compounds found after ozonation of natural waters resemble organics that can be isolated from natural waters without treatment (Glaze, 1986).

Ozonation of synthetic solutions of fulvic acid under simulated water treatment plant conditions and ozonation of natural, high-humic water from a creek were examined by Lawrence et al. (1980) and by Anderson et al. (1985). The predominant oxidation products included numerous alkyl phthalates, mono- and di-carboxylic aliphatic acids, and a few cyclic keto-compounds. Schalenkamp (1978) and Trussell (1985) have shown the presence of the aldehydes butanal, pentanal, hexanal, heptanal, octanal, nonanal, decanal, undecenal, dodecanal, tridecanal, and tetradecanal. Concentrations in ozonated drinking water have typically been 0.01 to 0.1 μg/L. At present, no work has been done with methods that would detect ketones and acids at these levels.

3.5 Effectiveness of Disinfection

Virtually all enteric pathogenic microorganisms (i.e., viruses, bacteria, and parasites found in the human intestinal tract) are killed or inactivated by chlorine and other disinfectants commonly used to treat drinking water. The amount of disinfectant required depends on such factors as water pH, temperature, and turbidity. The type of organism is also important. Viruses and protozoan parasites, such as *Giardia,* are more resistant than bacterial pathogens (Table 3-8).

Table 3-8.   Summary of Concentration Times (mg/L × Minutes) for 99% Inactivation of Various Microorganisms by Disinfectants at 5°C[a,b]

Micro-organism[b]	Free Chlorine (pH 6–7)	Preformed Chloramine (pH 8–9)	Chlorine Dioxide (pH 6–7)	Ozone (pH 6–7)
E. coli	0.034–0.05	95–180	0.4–0.75	0.02
Polio 1	1.1–2.5	768–3,740	0.2–6.7	0.1–0.2
Rotavirus	0.01–0.05	3,806–6,476	0.2–2.1	0.006–0.06
Phage f2	0.08–0.18	—	—	—
G. lamblia cysts	47–>150	—	—	0.5–0.6
G. muris cysts	30–630	—	7.2–18.5	1.8–2.0

[a]Hoff (1986).
[b]*E. coli* is a bacterium, *G. lamblia* and *G. muris* are protozoans, and the other microorganisms are viruses.

Filtration of surface waters aids the removal of pathogens (particularly the protozoan parasites), as well as turbidity, which interferes with disinfection. Filtration thus reduces the amount of disinfectant required but does not eliminate the need for disinfection. The surface water treatment requirements currently proposed by EPA for controlling waterborne disease outbreaks would require use of disinfection and filtration/disinfection processes that achieve a 3- to 4–log removal of *Giardia* or viruses (Regli, 1987; EPA, 1987). If implemented, these regulations will determine the future levels of disinfection in surface water supplies in the United States. They could significantly affect the type and dose of disinfectant used by utilities in this country.

3.5.1 Risks of Waterborne Infectious Disease

While disinfection and other treatment processes dramatically affect the incidence of certain enteric diseases, they do not necessarily eliminate these diseases. The effect of treatment is not equal for all enteric diseases (Table 3-9). In particular, some highly infectious parasites and viruses are more resistant to inactivation and removal by water treatment than other microbes (e.g., bacteria). (In the years prior to 1971, when viruses and parasites could not be readily identified, the bacterially transmitted illnesses typhoid and shigellosis were the leading known waterborne diseases.)

Table 3-9. Estimated Reductions in Waterborne Diseases Following Treatment of Water Supplies[a]

Disease	Estimated Reduction (%) In the Incidence as a Result of Improved Water Treatment[b]
Cholera	90
Typhoid	80
Leptospirosis	80
Viral hepatitis	10(?)
Enteroviruses	10(?)
Amoebic dysentery	50
Ascariasis	40
Legionella	—[c]

[a]Modified from Bradley (1977).
[b]The reductions were estimated by comparing the disease incidence during 1971 to 1977 with the incidence prior to improvements in drinking water treatment. Bradley (1977) compiled these figures using worldwide information (including U.S.). Water treatment improvements include disinfection and other typical treatment train technologies that are summarized in Figure 2-1.
[c]Legionella is primarily a problem in distribution systems rather than one of water sources.

Based on the most recently available data (Craun, 1986), the risk of acquiring a microbial-caused illness from water in the United States is approximately 4×10^{-5} per year, or 2.8×10^{-3} during an individual's lifetime (based on an average of 9,696 cases of illness between 1981 and 1983, a U.S. population of 240 million, and a 70-year lifetime). Since no reporting of waterborne disease outbreaks is required in the United States, the risk level is potentially greater because many outbreaks go unreported (Craun, 1986). These risk figures are largely developed from data on the consumption of untreated drinking water and are based only on data from outbreaks with a sufficient number of cases to be investigated and reported by public health agencies.

The effect of low levels of enteric pathogenic microorganisms in a water supply is more difficult to document. Infectious disease case studies with enteric viruses and parasites indicate that relatively low numbers of viruses and parasites, perhaps only 1 or 2 infectious units, are capable of causing infections (Ward and Akin, 1984; Rendtorff, 1954). Constant exposure to hepatitis A virus at concentrations of 1 infection unit per 10 L or less would eventually cause illness in every person consuming the water, with a risk of mortality of 0.6 percent (the current mortality rate observed in the United States) (Table 3-10). Cases of infectious hepatitis caused by a poor water supply are substantiated by the close association between antibodies to the disease and sanitary conditions in various countries of the world (Fresner, 1984). In developing countries hepatitis antibody prevalence exceeds 80 percent to 90 percent.

Table 3-10.   Mortality Rates for Enteric Bacteria and Viruses[a,b]

Organism		Mortality Rate (%)
Salmonella		0.2
Shigella		0.13
Legionella		—[c]
Hepatitis A		0.6
Coxsackie	A2	0.5
	A4	0.5
	A9	0.26
	A16	0.12
Coxsackie B		0.59–0.94
Echo	6	0.29
	9	0.27
Polio	1	0.9

[a]Assaad and Borecka (1977); Centers for Disease Control (1985); Berger (1986).
[b]Data for polio, coxsackie, and echo probably represent only hospitalized cases.
[c]Mortality data limited (see text).

The risks of infection, illness, and mortality to a susceptible population (i.e., with no preexisting protective antibodies) from low levels of enteric viruses and bacteria in drinking water can be estimated using available mortality data (Gerba and Haas, 1986). Figures 3–6 and 3–7 show estimates of risks from exposure to echovirus 6 and hepatitis A6. With enteric viruses, however, infection does not always result in clinical illness. Additionally, only broad estimates of risk can be made since the types and virulence of enteric organisms in a contaminated water supply probably vary greatly over any lengthy time period. Such estimates do, however, indicate that consumption of even low levels of pathogens may pose a significant risk of disease.

Legionella is a critically important disease organism because the entire U.S. population is at risk. The reasons for this are the ubiquitous occurrence of the organism in aquatic systems, and the lack of knowledge concerning the relative importance of sources and factors controlling human susceptibility. The source of most infections is not known, although water is usually strongly implicated.

Legionellosis has been reported in more than 26 countries. Case rates for 1976 to 1981 varied from 0.11 to 0.35 per 100,000 persons in the United States. As the disease has become more familiar, the reported case incidence has been increasing. An estimated 250,000 cases of pneumonia are caused by *Legionella* each year and approximately 800,000 cases of pneumonia are of unknown etiology (Fraser, 1980).

Infectivity data for humans are not available for *Legionella*. Data for guinea pigs suggest that the ID_{50} (i.e., the dose required to infect 50 percent of the exposed animals) is approximately 10 organisms. The LD_{50} (i.e., the dose required to kill 50 percent of the exposed animals) is approximately 10^2 to 10^5 organisms. Guinea pigs represent the most susceptible experimental animals relative to humans. Rats, which are more resistant relative to humans, showed only 11 percent mortality when exposed via the aerosol route. Most rats acquired mild symptoms compared to the guinea pigs (Davis et al., 1983).

Legionellae are abundant in ambient surface water, but some data suggest they are less prevalent or absent in ground water. The number of cases of disease attributable directly to drinking water is unknown.

Susceptibility to *Legionella* appears to increase with age. Also, immunocompromised individuals are highly susceptible. However, the infectious dose for humans, especially compromised individuals, is not known.

Risk estimates could be improved if data were available on the occurrence of enteric viruses and parasites in raw and drinking water supplies in the United States. Although these microorganisms are an increasingly well-documented cause of waterborne disease in the United States today, few systematic surveys have been attempted.

EPA proposed a Maximum Contaminant Level Goal (MCLG) for *Le-*

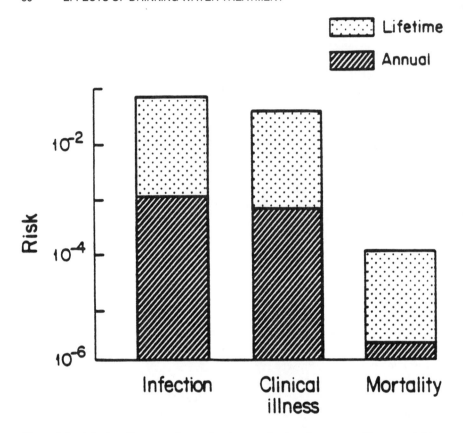

Figure 3-6. Infection, illness, and mortality due to echovirus 6 exposure. The annual risk of infection, illness, and mortality from echovirus 6 was estimated assuming consumption of water containing 1 viral infectious unit in 1,000 L and daily consumption of 2 L of drinking water per person per day. The minimum infectious dose data come from Schiff et al. (1984) for echovirus 12, and a post-ingestion probability is assumed to have a beta distribution (Haas, 1983). Clinical illness rates are from Feigin and Cherry (1981) and mortality from Table 3–12. Methods from Gerba and Haas (1986).

gionellae of zero since it takes only a few *Legionella* organisms (perhaps even a single organism) to pass through the treatment and distribution system and enter into an air conditioning system or plumbing system, proliferate under certain conditions, and possibly cause disease if a person, especially a compromised individual, is exposed via aerosols.

Although methods exist for recovering and enumerating *Legionellae,* it is not technically or economically feasible to monitor for these organisms. EPA therefore proposed a treatment technique rather than a Maximum Contaminant Level (MCL) for *Legionella.* In particular, EPA believes that filtration

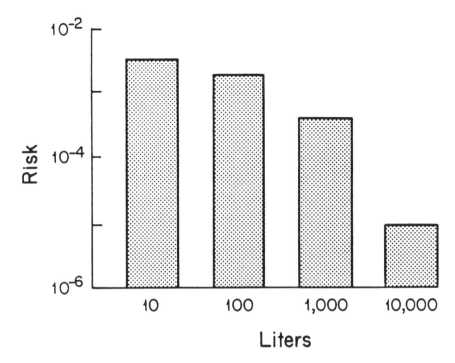

Figure 3-7.   Heptatis risk relative to water consumption. The estimated annual risk of mortality from exposure to hepatitis, given consumption of water containing one viral infectious unit in the volume indicated. For the purpose of this risk analysis, it was assumed that a person consumes 2 L of drinking water per day. The minimum infectious dose data come from Schiff et al. (1984) for echovirus 12, and a post-ingestion probability is assumed to have a beta distribution (Haas, 1983). Mortality from Table 3–12. Methods from Gerba and Haas (l986).

and disinfection would remove and/or inactivate *Legionella* in source waters, thereby reducing chances that *Legionella* will be transported through the system and that growth might occur in the distribution system or in hot water systems.

3.6 Health Risks Associated with Disinfection

This section examines the risks of adverse health effects from exposure to a disinfectant or its by-products. Possible health effects include carcinogenic, mutagenic, developmental, and reproductive effects as well as neurotoxicity and hepatotoxicity. In addition, Revis et al. (1986) reported that very low levels of chlorine, chlorine dioxide, and chloramine increase serum cholesterol and atherosclerotic lesions and depress thyroid hormones in pigeons. However, the National Research Council (NRC) Subcommittee on Drinking

Water Disinfectants was reluctant to apply these data to humans (NAS, 1987). The principal difficulties were (1) the suggestion by Revis et al. that these effects might be secondary to depressed thyroid function, and (2) the results of others (Bercz et al., 1982; Harrington et al., 1986) showing that chlorine dioxide was the only one of the three disinfectants that depressed thyroid function in mammalian species and that this effect occurred only at many times the concentrations used in the Revis et al. (1986) study.

3.6.1 Chlorine and Chlorine By-Products

Epidemiologic Data. Since 1974, when the use of chlorine as a disinfectant was shown to lead to the formation of trihalomethanes in finished drinking water (Bellar et al., 1974; Rook, 1974), a great deal of effort has gone into identifying other chlorine by-products and assessing the hazards these chemicals present to human health. According to epidemiological evidence, chlorination of drinking water may cause a slight increase in the risk of cancer. In particular, cancers of the bladder, colon, and rectum seem implicated (Craun, 1985).

The early studies in this area were reviewed by the National Research Council (NAS, 1980) and found to have a number of methodological problems—primarily lack of control over potentially confounding variables, small and variable increases in the relative risk, and inadequate documentation of exposure. However, more rigorous studies conducted since the NRC review (Cantor et al., 1985; Cragle et al., 1985; Young et al., 1987) confirm some of the observations in the early studies.

In the Cragle et al. (1985) case-comparison study of colon cancers and hospital-comparison subjects among North Carolina white residents, odds ratios of 1.38, 2.15, and 3.36 were observed for home consumption of chlorinated water for 16 or more years and colon cancer in 60-, 70-, and 80-year olds. These odds ratios suggest a weak-to-moderate association between water chlorination and colon cancer in the study population.

In an analysis of his 1978 epidemiological data, Cantor et al. (1987) found that, for one subgroup, i.e., those drinking greater than average amounts of water and exposed to chlorinated surface water for more than 40 years, there was an association (odds ratio of 3.1) with a small increased risk of bladder cancer.

Attempts to associate the development of cancer with specific chlorination by-products (e.g., trihalomethanes) have not been particularly successful (Young et al., 1987). This is not surprising given the large variety of disinfection by-products with carcinogenic and/or mutagenic properties that are known to be generated at small concentrations (Bull, 1986). Therefore, it is unlikely that one by-product would stand out as solely responsible for these small increases in cancer risk.

Animal Data. Table 3-11 summarizes the evidence concerning the mutagenicity and carcinogenicity of chlorine and its by-products. The table does not indicate the toxicological effects in question, but they include neurotoxicity, liver toxicity, and hematologic effects. In addition, the table lists the estimated cancer risk associated with the consumption of the contaminant in water. The data that support the numbers given in Table 3-11 are briefly discussed below. The carcinogenic risk attributable to some of these chlorination by-products would not be expected to exceed the estimates provided and may actually be considerably smaller for some chemicals. However, it should be noted that not all chlorination by-products have been toxicologically evaluated.

Sodium Hypochlorite. Three studies now indicate that sodium hypochlorite, one of the ionic forms of chlorine in solution, is not carcinogenic when supplied to animals for significant portions of their lifetimes. Rats provided with 100 mg/L sodium hypochlorite in drinking water for over seven generations did not show any indication of adverse effects or increased tumor incidence (Druckrey, 1968). Recently, Hasegawa et al. (1986) reported the results of studies in Fischer 344 rats maintained on 500 to 2,000 mg/L sodium hypochlorite for 104 weeks. No increases in tumor incidence were observed in either sex. Kurokawa et al. (1986) summarized these results and reported that mice given 500 or 1,000 mg/L of sodium hypochlorite for 103 weeks did not develop increased incidences of tumors at any organ site.

The National Toxicology Program (NTP) is presently evaluating the carcinogenic potential of hypochlorite. However, in the absence of data indicating that hypochlorite is carcinogenic, a safe level of 35 mg/L of sodium hypochlorite or 24.5 mg/L of hypochlorite ion can be derived based on the recent lifetime study of Hasegawa et al. (1986). The effect noted in this study was a slight increase in liver damage and depressed body weights at 2,000 mg/L. No adverse effects were observed at 1,000 mg sodium hypochlorite/L.

Major By-Products. There are three major groups of chlorination by-products: trihalomethanes, chlorinated acetic acids, and haloacetonitriles. The trihalomethanes require particular attention. There is also evidence that chlorine may produce a number of other by-products such as chlorophenols, chlorinated aldehydes, and ketones.

(a) Trihalomethanes. To date, three THMs have been implicated as carcinogens in animals: chloroform, chlorodibromomethane, and bromodichloromethane. A National Cancer Institute study (NCI, 1976) demonstrated that *chloroform* was capable of producing liver and kidney tumors in mice when administered in corn oil by stomach tube at high doses. A recent study was designed to specifically assess the risks associated with chloroform in drinking water and to extend these data to lower doses. This study confirmed the

Table 3-11.   Summary of Evidence Concerning Mutagenicity and Carcinogenicity of Chlorine and Its By-Products

Chemical	Qualitative Data			Quantitative Data	
	In vitro Mutagen- icity	In vivo Mutagen- icity	In vivo Carcino- genicity	Cancer Risk Based on Consump- tion of Water	10^{-6} Cancer Risk (μg/L)
Hypochlorite	+ (a)	—(b)	—(c)	NC(c,d)	
Trihalomethanes:					
Chloroform	—(e)	+ (f)	+ + (g)	6.1×10^{-3}	6
Chlorodibromomethane	+ (h)	—(h)	+ (i,j)		
Bromodichloromethane	+ (h,k)	—(h)	+ (j)		
Bromoform	+ (h,k)	—(h)	+ (j)		
Chlorinated Acids:					
Dichloroacetic	—(l)	ND(m)	+ (n)		
Trichloroacetic	—(l,o,p)	ND	+ (n,q)		
Chlorinated Aldehydes:					
Trichloroacetaldehyde	+ (r)	ND	+ (s)	NE(t)	
2-Chloropropenal	+ (u)	ND	ND	ND	
3,3-Dichloropropenal	+ (u)	ND	ND	ND	
2,3,3-Trichloropropenal	+ (u)	ND	ND	ND	
Haloacetonitriles:					
Dichloroacetonitrile	+ (k,v)	—(w)	—(v)	ND	
Dibromoacetonitrile	—(v)	—(w)	+ (v)	NE	
Bromochloroacetonitrile	+ (v)	—(w)	+ (v)	NE	
Trichloroacetonitrile	—(v)	—(w)	—(v)	ND	
Chlorophenols:					
2-Chlorophenol	ND	ND	ND	ND	
2,4-Dichlorophenol	—(x)	ND	ND	ND	
2,4,6-Trichlorophenol	± (y)	+ (y)	+ (y,z)	2×10^{-2}	175
Chlorinated Ketones:					
1,1-Dichloroacetone	(u)	ND	—(aa)	ND	
1,3-Dichloroacetone	(u)	ND	+ (bb)	ND	
1,1,1-Trichloroacetone	(u)	ND	—(aa)	ND	
1,1,3,3-Tetrachloro- acetone	(u)	ND	—(j)	ND	

[a]Rosenkrantz (1973) and Wlodkowski and Rosenkrantz (1975).
[b]Meier and Bull (1985).
[c]Carcinogenesis bioassays of chlorine have been conducted by Druckrey (1968), Hasegawa et al. (1986), and Kurokawa et al. (1986) without evidence of carcinogenic responses.
[d]NC indicates that appropriate long-term carcinogenesis bioassays have been conducted without evidence of carcinogenic effects.
[e]Reviewed by Bull (1986).
[f]Morimoto and Koizumi (1983).
[g]NCI (1976); Jorgenson et al. (1985); Roe et al. (1979).
[h]Ishidate et al. (1982).
[i]Dunnick et al. (1985).
[j]Theiss et al. (1977); Tumasonis et al. (1985).

[k]Simmon et al. (1977).
[l]Waskell (1978).
[m]ND indicates that no studies were identified that addressed the effect.
[n]Herren-Freund et al. (1987).
[o]Nestmann et al. (1980).
[p]Rapson et al. (1980).
[q]Parnell et al. (1986).
[r]Bignami et al. (1980).
[s]Rijhsingham et al. (1986).
[t]NE indicates that available data suggest that the chemical carcinogenic properties are not appropriate for making risk estimates.
[u]Meier et al. (1985a).
[v]Bull et al. (1985).
[w]Meier et al. (1985b).
[x]Rasanen et al. (1977).
[y]Fahrig et al. (1978).
[z]NCI (1979).
[aa]Bull and Robinson (1985).
[bb]Robinson et al. (1986).

NCI data on rat kidney tumors but found that chloroform in drinking water did not induce liver tumors in mice (Jorgenson et al., 1985). Subsequent experiments indicated that liver damage induced in mice by chloroform was significantly enhanced by a corn oil vehicle (Bull et al., 1986). Liver tumors can be produced in rats administered chloroform, but only if the treatment is preceded by treatment with diethylnitrosamine, a strong carcinogen (Deml and Oesterle, 1985). These data suggest that the kidney tumors in rats exposed to chloroform in drinking water are more appropriate than the liver data for estimating carcinogenic risks from drinking water exposure (NAS, 1987). Total kidney tumors from the Jorgenson study were the basis for the cancer risk estimate in Table 3-11.

(b) Chlorodibromomethane was reported to increase the incidence of liver tumors in B6C3F1 mice but produced no evidence of carcinogenic activity in rats (Dunnick et al., 1985).

Tumasonis et al. (1985) found that female Wistar rats treated with *bromodichloromethane* had increased incidences of hepatic adenofibrosis and neoplastic nodules. More recently, the National Toxicology Program's bioassay of bromodichloromethane (NTP, 1986) revealed that the chemical in a corn oil vehicle can induce tumors in different species and in different organs. It produced intestinal and kidney tumors in both male and female rats, kidney tumors in male mice (a rare tumor in mice); and liver tumors in female mice. EPA has not yet quantified the carcinogenic risk for bromodichloromethane.

Theiss et al. (1977) found that *bromoform* increased the incidence of lung adenomas in strain A mice. Since adenomas are benign tumors and occur spontaneously at a high incidence in this strain, they can only be taken as suggestive of a carcinogenic effect rather than as definitive proof of carcinogenesis.

(c) Chlorinated Acetic Acids. In the second major group of chlorinated by-products, the chlorinated acetic acids, dichloroacetic acid has been studied more thoroughly than trichloroacetic acid because of its therapeutic use as a hypoglycemic agent. In subchronic studies, *dichloroacetic acid* produced hindlimb weakness, vacuolation of myelinated tracts in the brain, and degeneration of the germinal epithelium of the testis as well as a variety of other toxic effects in rats and dogs (Katz et al., 1981). A reversible polyneuropathy had been previously reported in a 21-year-old man who had been administered dichloroacetic acid at a dose of about 50 mg/kg for 16 weeks (Stacpoole et al., 1979). Based on these and other data, the National Research Council suggested that the levels of dichloroacetic and trichloroacetic acids should not exceed 0.12 and 0.05 mg/L respectively in drinking water (NAS, 1987).

A recent study found that dichloroacetic and trichloroacetic acids induce a high incidence of liver tumors in mice in a relatively short time (Herren-Freund et al., 1987). The carcinogenic activity of these chemicals needs to be confirmed in a second species.

(d) Haloacetonitriles. In the third class of chlorination by-products—the haloacetonitriles—dichloroacetonitrile, dibromoacetonitrile, and bromochloroacetonitrile (these three haloacetonitriles are also known as dihaloacetonitriles) have been identified in drinking water in the low μg/L range (Reding et al., 1986). Trichloroacetonitrile has not recently been found in drinking water supplies. Little toxicological information is available on these compounds. Experimental animals tolerate doses up to 45 mg/kg per day with little difficulty (Hayes et al., 1986). Maternally toxic doses (55 mg/kg per day) also cause toxic effects in the fetus. Based on these data, the National Research Council recommended that *dichloroacetonitrile* be limited to 0.056 mg/L and *dibromoacetonitrile* not exceed 0.023 mg/L (NAS, 1987). The NAS assumed a relative source contribution of 20 percent via drinking water. Since these substances are considered unique to drinking water, the actual relative source contribution may be much higher (90 percent to 100 percent). Because these chemicals may be mutagenic and carcinogenic, the National Research Council committee strongly recommended that the database be more completely developed.

Dichloroacetonitrile and bromochloroacetonitrile are mutagenic in bacteria (Bull et al., 1985). All three dihaloacetonitriles can increase the rate of sister chromatid exchanges in mammalian cells in vitro. Dibromoacetonitrile and bromochloroacetonitrile also initiate skin tumors in mice (Bull and Robinson, 1985). The possible carcinogenic activity of dihaloacetonitriles should be evaluated in lifetime studies in experimental animals.

Other By-Products. In addition to the major groups of by-products, a variety of other chemicals are formed from reactions of chlorine with fulvic and humic acids—the natural constituents of most surface waters. No studies have been conducted to determine whether these compounds appear at significant concentrations in finished drinking water. Many of the identified chemicals are mutagenic in cell systems (Bull, 1986). Toxicological information is insufficient to estimate safe levels of these substances in drinking water. In general, it is unlikely that any of these chemicals would exceed a concentration of few µg/L in drinking water and most probably occur at lower concentrations. The best known of the other by-products that may have toxic properties of concern are chlorophenols, chlorinated ketones, and chlorinated aldehydes. These are included in Table 3–11, with an indication of the data that are presently available.

Mixtures. A number of studies have attempted to document the hazards of mixtures of chlorinated by-products rather than of individual compounds. Many of these studies assessed whether chlorination increased mutagenic activity of organic material in bacteria. In some studies, the test material consisted of a concentrate of chlorinated drinking water; other studies used surrogate substrates such as humic acids at high concentrations so that no concentration was necessary. These studies consistently show that the chlorination of drinking water or of substrates similar to those found in drinking water increases the level of mutagenic activity detectable in bacterial systems and other in vitro systems (Bull, 1986). However, attempts to demonstrate mutagenic or carcinogenic activity in vivo with such mixtures at levels of up to 4,000 times that encountered in drinking water have been uniformly negative (Meier and Bull, 1985; Miller et al., 1986; Van Duuren et al., 1986). While these data do not prove that chlorination of drinking water fails to increase carcinogenic and mutagenic risks, they do indicate that the risks are probably not large.

3.6.2 Chloramines

In general, chloramination produces fewer by-products than chlorination. No evidence suggests that chloramination produces by-products that are not produced by chlorine; rather it produces similar by-products at lower concentrations. Theoretically, however, chloramines could produce different by-products, such as organic chloramines, nitriles, and compounds with chlorine and amino groups across double bonds (Rice and Gomez-Taylor, 1986). Since chlorine forms chloramines when ammonia is present in the source water, these differences may be of little concern because the same compounds would be present with chlorination as well.

The health hazards of chloramine differ from those of chlorine only to the extent that chloramine itself may present a hazard; however, very little work has been done to evaluate this issue (Table 3-12). The NTP is currently conducting a carcinogenicity bioassay for chloramine. For noncancer effects, the most significant documented toxicity for chloramine is liver damage, increased mitotic figures, and unusual chromatin patterns in mice exposed to 100 mg/L or more of chloramine. Based on these data, the NRC Safe Drinking Water Subcommittee on Disinfectants recommended that chloramine levels in drinking water should not exceed 0.166 mg/L for a 10-kg child. A level of 0.58 mg/L was recommended for a 70-kg adult. The NAS assumed a relative source contribution of 20 percent via drinking water. Since these substances are considered unique to drinking water, the actual relative source contribution may be much higher (90 percent to 100 percent).

Table 3-12. Summary of Data on Mutagenic or Carcinogenic Activity of Chloramine, Chlorine Dioxide, Chlorite, and Chlorate

Compound	In vitro Mutagenicity	In vivo Mutagenicity	In vivo Carcinogenicity
Chloramine (mono)	+[a]	_[b]	ND[c]
Chlorine Dioxide	ND	_[b]	ND
Chlorite	ND	_[b]	±[d]
Chlorate	+[e]	_[b]	ND

[a]Shih and Lederberg (1976).
[b]Meier and Bull (1985).
[c]ND indicates that no appropriate study was identified that addressed the end point.
[d]Kurokawa et al. (1986).
[e]Eckhardt et al. (1982).

A typical range of chloramine concentrations in drinking water supplies where it is used as a primary disinfectant or to provide a residual in the distribution system is 1.5 to 2.5 mg/L—a substantially higher level than the NRC-recommended level. However, a concentration of 0.8 mg/L used as a secondary disinfectant might be effective in preventing growth in a distribution system. One possibility for accomplishing effective disinfection while still minimizing the formation of by-products was recommended by NRC (NAS, 1980): add ammonia after a sufficient contact time with chlorine has passed.

3.6.3 Chlorine Dioxide and Its By-Products

Considerable interest has developed in using chlorine dioxide as a primary disinfectant for drinking water. It is a very effective disinfectant and a residual can be maintained to control infestations of the distribution system. Until recently very little dependable toxicological information existed for this compound, despite its use in drinking water treatment to control undesirable tastes and odors. The higher doses involved in the chemical's use as a primary disinfectant and the recognized toxicity of two of its by-products, chlorite and chlorate, have sparked diverse research efforts that give at least a partial view of the hazards associated with this compound. At present, only 22 of 907 water supplies surveyed by the American Water Works Association Research Foundation reported the use of chlorine dioxide (AWWARF, 1987). However, anticipated reduction in the standards for trihalomethanes may force a number of systems to consider chlorine dioxide as a primary disinfectant in the future.

A two-year study of the gross effects of chlorine dioxide in drinking water demonstrated that concentrations of 100 mg/L significantly decreased survival of both male and female rats (Haag, 1949). More recent studies found that administration of chlorine dioxide at these same levels led to depressed thyroid function in African green monkeys (Bercz et al., 1982) and in neonatal rats (Orme et al., 1985). Neonatal rats subjected to doses of 14 mg/kg body weight by stomach tube (equivalent to 100 mg/L of drinking water for the rat) showed delayed brain development and altered behavioral activity as young adult animals, even though treatment was suspended when the rat pups were weaned (Taylor and Pfohl, 1985).

Based on these studies, the NRC Subcommittee on Drinking Water Disinfectants (NAS, 1987) recommended that the levels of chlorine dioxide in drinking water should not exceed 0.06 mg/L, assuming the consumption of 1 L of drinking water per day by a 10-kg child (modified thyroid function has more serious implications during development). The NAS-recommended safe level for adults is 0.2 mg/L assuming consumption of 2 L of drinking water per day by a 70-kg adult. The NAS assumed a relative source contribution of 20 percent via drinking water. Since this substance is considered unique to drinking water, the actual relative source contribution may be much higher (90 percent to 100 percent).

By-Products. The use of chlorine dioxide as a primary disinfectant is further complicated by the acute toxicological effects of its by-products, chlorite and chlorate. Both compounds produce methemoglobinemia. More recent investigations have demonstrated that chlorite produces an hemolytic anemia at doses below those that result in methemoglobin formation (Heffernan et al., 1979). Relatively large subgroups of the American population are more susceptible to such damage, which is an additional cause for concern.

The mutagenic and carcinogenic hazards of chlorate and chlorite have not been clearly established. Chlorate was reported mutagenic in bacteria and *Drosophila* (Eckhardt et al., 1982). Other investigators have not demonstrated mutagenic effects of either compound in vivo (Meier et al., 1985a). Sodium chlorite did increase hyperplastic nodules in the liver and adenomas in the lung of male B6C3F1 mice, but not in females (Kurokawa et al., 1986). Although the hepatocellular carcinoma incidence was increased in these animals as well, the increase was not statistically significant. Chlorite has also been reported to promote skin tumors in mice, but these data came from a study involving limited numbers of animals (Kurokawa et al., 1984).

Based on the results of a human study (Lubbers et al., 1981) in which no adverse effects were observed at 0.034 mg/kg/day, a level of 0.007 mg/L was recommended by the NRC Subcommittee on Drinking Water Disinfectants for both chlorite and chlorate (NAS, 1987). The NAS assumed a relative source contribution of 20 percent via drinking water. Since these substances are considered unique to drinking water, the actual relative source contribution may be much higher (90 percent to 100 percent) and, therefore, the NAS assumption may underestimate the relative source contribution. The NAS level was based on the fact that children are the segment of the population most sensitive to hemolytic agents.

The prior level recommended by NRC was 0.21 mg/L (assuming 100 percent of the exposure comes from drinking water), based on effects in cats (NAS, 1980). Since the human study failed to demonstrate a positive effect, it may be more appropriate to use the data from cats for estimating the reference dose. This figure, however, should be modified to take into account children's higher sensitivity to oxidants than adults. Applying conventional calculations of acceptable exposure to the data from cats, 0.06 mg/L is acceptable as the safe level for a 10-kg child who drinks 1 L of water per day if drinking water is the only source of exposure. (The NAS assumed a relative source contribution of 20 percent via drinking water. Since these substances are considered unique to drinking water, the actual relative source contribution may be much higher [90 percent to 100 percent]). Most current methods of applying chlorine dioxide as a disinfectant would result in considerably higher concentrations of chlorite in the water distributed to the consumer.

Little information is available concerning organic by-products that might be produced by chlorine dioxide. If pure chlorine dioxide is utilized for disinfection, trihalomethanes are not formed (NAS, 1980). There is evidence to suggest that application of chlorine dioxide may produce some chlorine, but this has not been clearly demonstrated (Rice and Gomez-Taylor, 1986). Chlorine dioxide does produce oxidized polar species. Because these polar by-products are difficult to detect, few efforts have been made to characterize them. No attempts have been made to survey drinking waters to determine the relative concentrations that are produced. Consequently, there is virtually no information appropriate for evaluating the hazards that might be associated with organic by-products of chlorine dioxide use in drinking water disinfection.

3.6.4 Ozone

Ozone is a very effective disinfectant. Its major disadvantage is its instability; residual levels cannot be maintained in water distribution systems to control microbe infestations. From a toxicological perspective, the lack of a residual at the tap means no toxicological hazards can be attributed to ozone itself. Ozone is now used in over 1,000 drinking water plants in Europe. As with the other reactive chemicals used in drinking water disinfection, ozone reacts with and changes the nature of the organic material that is present. Ozone is very reactive and undoubtedly produces an array of by-products (Rice and Gomez-Taylor, 1986). The by-products of potential toxicological significance that have been discovered are aldehydes (butanal, pentanal, hexanal, n-heptanal, octanal, and n-nonanal, decanal, undecenal, dodecanal, tridecanal, and tetradecanal) from the water works in Zurich, Switzerland (NAS, 1980) and as reported in studies of Schalenkamp (1978) and Trussell (1985). However, the health effects significance of these compounds has not been determined.

The potential for increased atmospheric ozone due to ozonation has not been examined. However, ozone used in drinking water disinfection is administered in a closed system which is additionally equipped with a catalytic or thermal system to remove ozone from any off gases. Therefore, it is unlikely that ozone would be released to the atmosphere.

3.7 Comparison of the Risks and Benefits of Disinfection

Section 3.5 estimated the risk of contracting infectious disease if disinfection is not practiced, and Section 3.6 estimated the toxicological risks associated with the use of disinfectants. These estimates are, in fact, gross oversimplifications. Disinfection is not the only water treatment that reduces microbial agents in finished drinking water. Filtration, for example, significantly reduces microbes in contaminated water. Consequently, if filtration is applied, the risks attributed to nondisinfected water may be somewhat exaggerated. Similarly, many of the toxicological effects attributed to disinfectants can be considerably reduced if waters are treated (e.g., by granular-activated carbon) to reduce the organic precursors in the source water.

The toxicological risks introduced through the use of different disinfectants vary widely. Under standard approaches to calculating safe levels, hypochlorite—the aqueous form of chlorine at neutral pH—can be present at the tap at 24.5 mg/L. This level is many times the concentration required for primary disinfection of most source waters or for the chemical's effectiveness as a residual in maintaining microbiological water quality in the distribution system. Conversely, the levels of chloramine and chlorine dioxide required to act as residuals in the distribution system approach or even exceed (depending on residence time in the system before the first withdrawal at a tap) those which would be considered safe. No limits have been developed for ozone because of its instability in water.

When the by-products of disinfection are considered, questions of safety become much more complex, although the theoretical risks are small. The amount and nature of these by-products depend not only on the disinfectant used but also on the type of precursors in the source water and on water conditions such as temperature and pH. A clear example of this interdependency is the wide variation among water systems in by-products that are formed during *chlorination*. Dichloroacetic acid, trichloroacetic acid, trichloroacetaldehyde, and bromoform are frequently occurring by-products. A small increase in the carcinogenic risk is introduced by two by-products of chlorination.

While *chlorination* of drinking water may introduce a carcinogenic risk, data are insufficient at this time to quantify this risk. Only chloroform and 2,4,6-trichlorophenol have been quantitatively assessed for possible human carcinogenicity. Research is in progress to determine the carcinogenic potential of several other chlorination by-products.

Following the use of chlorine as the primary disinfectant, *ammonia* is commonly added to maintain a more stable residual level of disinfectant in the distribution system.

Chloramination apparently produces by-products that are similar to those of chlorine, but at much lower concentrations (NAS, 1987). Consequently, the risks associated with by-products of chloramination in a particular system are less than with chlorine in the same water supply. The use of *chlorine dioxide* at concentrations that are effective in primary disinfection introduces chlorite as a by-product in virtually any water to which it is applied. In waters that contain high levels of natural organic matter, this problem can be considerably exacerbated. Chloramine and chlorine dioxide have thus not been demonstrated to be effective as primary disinfectants within the limitations that would have to be placed on their residual concentrations or those of their by-products at the consumer's tap.

Little concrete evidence appears to question the use of *ozone* in primary disinfection. Ozone is a very effective disinfectant, but because no residual of ozone can be maintained in a distribution system, a second disinfectant might be needed to prevent regrowth of microbes. Ozone is very reactive and produces a number of potentially significant by-products such as aldehydes. The health significance of these compounds has not been determined.

Risks from inadequate disinfection of drinking water are difficult to determine. In principle, estimates can be developed in a manner that is analogous to that used for chemical carcinogens. A certain dose (i.e., amount) of the infectious agent is required to initiate an infection. Those who contract the disease have a certain probability of dying from the disease. And any increase in infectious disease incidence within a population increases the likelihood of disease transmission by other means (i.e., interpersonal contact and through food). Disinfection of drinking water reduces the likelihood that infectious disease will be transmitted.

Properly applied disinfection provides an effective barrier against waterborne infectious disease. In many water systems in the United States, it is the only barrier. If disinfection of a water system that uses a contaminated source and that currently depends entirely on disinfection were abandoned, everyone within the community would contract one or more waterborne diseases during their lifetimes (i.e., 100 percent infection). Under such circumstances, the number of deaths from these diseases would equal the mortality rate for each disease. For example, 0.13 percent of individuals infected with *Shigella* would die while 0.44 percent of individuals with hepatitis A would die. If competing causes of death are not considered, the overall risk of death can be obtained by multiplying the mortality rates of each infectious agent that can be transmitted via drinking water. Therefore, lifetime risks for death by infectious disease in such situations could easily approach those seen in countries lacking adequate disinfection and/or other treatment technologies.

It is much more difficult to estimate the health benefits of applying disinfection to source waters lacking microbial contamination than to water contaminated with microbes. In such cases, the probability of accidental contamination of the source or distribution system would be the primary concern. The data needed for such assessments, however, are not available. The question is, how much more frequently would such problems occur if the water source of a community is not disinfected? For example, accidental cross-connections between sewage and drinking water lines occur frequently. The lack of residual disinfectant in sewage systems greatly increases the likelihood of colonization of the distribution system and could result in large-scale waterborne outbreaks.

Despite the overwhelming importance of adequate disinfection, the possible health risks that arise from the use of disinfectants must not be neglected. Efforts should always be directed at minimizing the overall health risks associated with drinking water. Formation of disinfection by-products can be minimized in many ways, including removing precursors, modifying water conditions (e.g., pH), closely controlling the amount of disinfectant which is added and the amount of residual, and removing the by-products themselves. These are all options that have received much less attention than has the search for disinfectants that do not produce undesirable by-products.

Perhaps the most important point to remember is to avoid overtreatment. Too frequently the response to problems in a water treatment plant is to add more disinfectant, but it should be clear now that such practices diminish—rather than improve—the quality of the water from a human health point of view.

REFERENCES

Anderson, L.J., et al. 1985. The reaction of ozone with isolated aquatic fulvic acid. *Org. Geochem.* 81:65–69.

APHA (American Public Health Association). 1985. Standard methods for the examination of water and wastewater. Greenberg, A., L. Clesceri, R. Trussell (eds.). New York, New York: American Public Health Association.

Assaad, F. and I. Borecka. 1977. Nine-year study of World Health Organization virus reports on fatal virus infections. *Bull. W.H.O.* 55:445–453.

AWWARF (American Water Works Association Research Foundation). 1987. National trihalomethane survey report. Prepared by Decision Research under the supervision of the Metropolitan Water District of Southern California.

Bellar, T.A., J.J. Lichtenberg, R.C. Kroner. 1974. The occurrence of organohalides in chlorinated drinking waters. *J. Am. Water Works Assoc.* 66:703–706.

Bercz, J.P., L. Jones, L. Garner, D. Murray, D.A. Ludwig, J. Boston. 1982. Subchronic toxicity of chlorine dioxide and related compounds in drinking water in the nonhuman primate. *Environ. Hlth. Perspect.* 46:47–55.

Berger, P. 1986. Personal Communication (Letter, June 4) to Charles P. Gerba.

Bignami, M., G. Conti, L. Conti, R. Crebelli, F. Misuraca, A.M. Puglia, R. Randazzo, G. Sciandrella, A. Carere. 1980. Mutagenicity of halogenated aliphatic hydrocarbons in *Salmonella typhimurium, Streptomyces coelicolor* and *Aspergillus nidulans*. *Chem. Biol. Interact.* 30:9–23.

Bradley, D.J. 1977. Health aspects of water supplies in tropical countries. *In:* Feachem, R., M. McGarry, M. Duncan (eds.), Water wastes and health in hot climates. New York: John Wiley & Sons, Inc., pp. 3–17.

Bull, R.J. 1986. Carcinogenic hazards associated with the chlorination of drinking water. *In:* Ram, N.M., E. Calabrese, R.F. Christman (eds.), Organic carcinogens in drinking water. New York: John Wiley & Sons, Inc., pp. 353–371.

Bull, R.J., J.M. Brown, E.A. Meierhenry, T.A. Jorgenson, M. Robinson, J.A. Stober. 1986. Enhancement of the hepatotoxicity of chloroform in B6C3Fl mice by corn oil: Implications for chloroform carcinogenesis. *Environ. Hlth. Perspect.* 69:49–58.

Bull, R.J., J.R. Meier, M. Robinson, H.P. Ringhand, R.D. Laurie, J.A. Stober. 1985. Evaluation of mutagenic and carcinogenic properties of brominated and chlorinated acetonitriles: By-products of chlorination. *Fund. Appl. Toxicol.* 5:1065–1074.

Bull, R.J. and M. Robinson. 1985. Carcinogenic activity of haloacetonitrile and haloacetone derivatives in mouse skin and lung. *In:* Jolley, R.L. et al. (eds.), Water chlorination: Chemistry, environmental impact and health effects, vol. 5. Chelsea, Michigan: Lewis Publishers, Inc., pp. 221–227.

Cantor, K.P., et al. 1987. Bladder cancer, drinking water source, and tap water consumption: A case-control study. *J. Natl. Cancer Inst.* December.

Cantor, K.P., R. Hoover, P. Hartge, T.J. Mason, D.T. Silverman, L.I. Levin. 1985. Drinking water source and risk of bladder cancer: A case-control study. *In:* Jolley, R.L. et al. (eds.), Water chlorination: Chemistry, environmental impact and health effects, vol. 5. Chelsea, Michigan: Lewis Publishers, Inc., pp. 145–152.

Centers for Disease Control. 1985. Hepatitis surveillance. Report no. 49. Atlanta, Georgia: Centers for Disease Control.

Colclough, C.A., et al. 1985. Organic reaction products of chlorine dioxide and natural aquatic fulvic acids. *In:* Jolley, R.L., et al. (eds.), Water chlorination: Chemistry, environmental impact and health effects, vol. 5. Chelsea, Michigan: Lewis Publishers, Inc.

Cotruvo, J.A. 1983. Introductory Remarks. *In:* Berger, P.S. and Y. Argaman (eds.) Assessment of Microbiology and Turbidity Standards for Drinking Water—Proceedings of a Workshop. December 2–4, 1981. Washington, DC: U.S. EPA Office of Drinking Water.

Cragle, D.L., C.M. Shy, R.J. Struba, E.J. Siff. 1985. A case-control study of colon cancer and water chlorination in North Carolina. *In:* Jolley, R.L. et al. (eds.), Water chlorination: Chemistry, environmental impact and health effects, vol. 5. Chelsea, Michigan: Lewis Publishers, Inc., pp. 153–159.

Craun, G.F. 1986. Waterborne diseases in the United States. Boca Raton, Florida: CRC Press.

Craun, G.F. 1985. Epidemiologic considerations for evaluating associations between the disinfection of drinking water and cancer in humans. *In:* Jolley, R.L. et al. (eds.), Water chlorination: Chemistry, environmental impact and health effects, vol. 5. Chelsea, Michigan: Lewis Publishers, Inc., pp. 133–143.

Croue, J.P., et al. 1987. Effect of pre-ozonation on the organic halide formation potential of an aquatic fulvic acid. Submitted to *Environ. Sci. Technol.* May.

Davis, G.S., W.C. Winn, G.W. Gump, J.F. Craighead, H.N. Beaty. 1983. Legionnaires' Pneumonia After Aerosol Exposure in Guinea Pigs and Rats. *Amer. Rev. Respir. Dis.* 126:1050–1057.

Deml, E. and D. Oesterle. 1985. Dose-dependent promoting activity of chloroform in rat liver foci bioassay. *Cancer Letters* 29:59–63.

Druckrey, H. 1968. Chlorinated drinking water toxicity tests involving seven generations of rats. *Food Cosmet. Toxicol.* 6:147–154.

Dunnick, J.K., J.K. Haseman, H.S. Lilja, S. Wyand. 1985. Toxicity and carcinogenicity of chlorodibromomethane in Fischer 344/N rats and B6C3Fl mice. *Fund. Appl. Toxicol.* 5:1128–1136.

Eckhardt, K., E. Gocke, M.T. King, D. Wild. 1982. Mutagenic activity of chlorate, bromate, and iodate. *Mutat. Res.* 97:85. (Abstract 42).

EPA. 1987. National Primary Drinking Water Regulations: Proposed surface water treatment requirements. *Fed. Regis.* 52:42178. November 3.

Fahrig, R., C.A. Nilsson, C. Rappe. 1978. Genetic activity of chlorophenols and chlorophenol impurities. *In:* Rao, K.R. (ed.), Pentachlorophenol: Chemistry, pharmacology, and environmental toxicology. New York: Plenum Press.

Feigin, R.D. and J.D. Cherry. 1981. Textbook of pediatric infectious diseases. Philadelphia, Pennsylvania: W.B. Saunders Co.

Fraser, D.W. 1980. Legionnaires' Disease: Four Summers' Harvest. *Am. J. Med.* 68:1–2.

Fresner, G. 1984. Hepatitis A. *In:* Belshe, R.B. (ed.), Textbook of human virology. Littleton, Massachusetts: PSG Publishing Company, pp. 707–727.

Gerba, C.P. and C.N. Haas. 1986. Risks associated with enteric viruses in drinking water. *In:* Janauer, G.E. (ed.), Progress in chemical disinfection. State University of New York, Binghamton, New York, pp. 460–468.

Glaze, W.H. 1986. Reaction products of ozone: A review. *Environ. Hlth. Perspect.* 69:151–157.

Greenberg, K. 1986. Letter (April 15, 1986) to utilities participating in 10-city survey, with attachment by P. Fair and R. Reding.

Haag, H.B. 1949. The effect on rats of chronic administration of sodium chlorite and chlorine dioxide in the drinking water. Report submitted to The Mathieson Alkali Works, Niagara Falls, New York. (Prepared by the Department of Physiology and Pharmacology, Medical College of Virginia, Richmond, Virginia.)

Haas, C.N. 1983. Estimation of risk due to low doses of microorganisms:

A comparison of alternative methodologies. *Am. J. Epidemiol.* 118:573–582.

Harrington, R.M., H.G. Shertzer, J.P. Bercz. 1986. Effects of chlorine dioxide on thyroid function in the African green monkey and the rat. *J. Toxicol. Environ. Health* 19:235–242.

Hasegawa, R., M. Takahashi, T. Kokubo, F. Furukawa, K. Toyoda, H. Sato, Y. Kurokawa, Y. Hayashi. 1986. Carcinogenicity study of sodium hypochlorite in F344 rats. *Fd. Chem. Toxic.* 24:1295–1302.

Hayes, J.R., L.W. Condie, J.F. Borzelleca. 1986. Toxicology of haloacetonitriles. *Environ. Hlth. Perspect.* 69:183–202.

Heffernan, W.P., C. Guion, R.J. Bull. 1979. Oxidative damage to the erythrocyte induced by sodium chlorite, *in vivo*. *J. Environ. Pathol. Toxicol.* 2:1487–1499.

Helms, C.M., R.M. Massanari, R. Zeitler, et al. 1983. Legionaires' disease associated with a hospital water system—a cluster of twenty four nosocomial cases. *An. Int. Med.* 9:172–178.

Herren-Freund, S.L., M.A. Pereira, G. Olson. 1987. The carcinogenicity of trichloroethylene and its metabolites, trichloroacetic acid and dichloroacetic acid, in mouse liver. *Toxicol. Appl. Pharmacol.* 90:183–189.

Hoff, J.C. 1986. Inactivation of microbial agents by chemical disinfectants. EPA Report no. 600/2–86/067. Cincinnati, OH: U.S. Environmental Protection Agency, Water Engineering Research Lab.

Ishidate, M., T. Sofuni, K. Yoshikawa, M. Hayashi. 1982. Studies on the mutagenicity of low boiling organohalogen compounds. Submitted to the National Institute of Hygienic Science. Tokyo, Japan: Tokyo Medical and Dental University.

Jorgenson, T.A., E.F. Meierhenry, C.J. Rushbrook, R.J. Bull, M. Robinson. 1985. Carcinogenicity of chloroform in drinking water to male Osborne-Mendel rats and female B6C3Fl mice. *Fund. Appl. Toxicol.* 5:760–769.

Katz, R., C.N. Tai, R.M. Diener, R.F. McConnell, D.E. Semonick. 1981. Dichloroacetate sodium: 3-month oral toxicity studies in rats and dogs. *Toxicol. Appl. Pharmacol.* 57:273–287.

Kurokawa, Y., S. Takayama, Y. Konishi, Y. Hiasa, S. Asahina, M. Takahasi, A. Maekawa, Y. Hayashi. 1986. Long-term *in vivo* carcinogenicity tests of potassium bromate, sodium hypochlorite, and sodium chlorite conducted in Japan. *Environ. Hlth. Perspect.* 69:221–235.

Kurokawa, Y., N. Takamura, Y. Matsushima, T. Imazawas, Y. Hayashi. 1984. Studies on the promoting and complete carcinogenic activities of some oxidizing chemicals in skin carcinogenesis. *Cancer Letters* 24:299–304.

Lawrence, J., et al. 1980. The ozonation of natural waters: product identification. *Ozone: Sci. Engin.* 2:55–64.

Lubbers, J.R., S. Chauhan, J.R. Bianchine. 1981. Controlled clinical evaluations of chlorine dioxide, chlorite and chlorate in man. *Environ. Hlth. Perspect.* 46:57–62.

McGuire, M. and R. Meadows. 1987. American Water Works Association Research Foundation trihalomethane survey—a progress report. Presented at the Conference on Current Research in Drinking Water Treatment, March 24–26. Cincinnati, Ohio.

Meier, J.R., H.P. Ringhand, W.E. Coleman, J.W. Much, R.P. Streicher, W.H. Kaylor, K.M. Schenk. 1985a. Identification of mutagenic compounds formed during chlorination of humic acid. *Mut. Res.* 157:111–122.

Meier, J.R., R.J. Bull, J.A. Stober, and M.C. Cimino. 1985b. Evaluation of chemicals used for drinking water disinfection for production of chromosomal damage and spermhead abnormalities in mice. *Environ. Mut.* 7:201–211.

Meier, J.R. and R.J. Bull. 1985. Mutagenic properties of drinking water disinfectants and by-products. *In:* Jolley, R.L. et al. (eds.), Water chlorination: Chemistry, environmental impact and health effects, vol. 5. Chelsea, Michigan: Lewis Publishers, Inc., pp. 207–220.

Miller, R.G., F.C. Kopfler, L.W. Condie, M.A. Pereira, J.R. Meier, H.P. Ringhand, M. Robinson, B.C. Casto. 1986. Results of toxicological testing of Jefferson Parish pilot plant samples. *Environ. Hlth Persp.* 69:129–139.

Morimoto K. and A. Koizumi. 1983. Trihalomethanes induce sister chromatid exchange in human lymphocytes *in vitro* and mouse bone marrow cells *in vivo*. *Environ. Res.* 32:72–79.

NAS (National Academy of Sciences). 1987. Drinking water and health, vol. 7. Washington, D.C.: National Academy of Sciences. In press.

NAS (National Academy of Sciences). 1980. Drinking water and health, vol. 2. Washington, D.C.: National Academy of Sciences.

NCI (National Cancer Institute). 1979. Bioassay of 2,4,6-trichlorophenol for possible carcinogenicity. National Cancer Institute. NCI-CG-TR-155.

NCI (National Cancer Institute). 1976. Report on the carcinogenesis bioassay of chloroform. Bethesda, Maryland: National Cancer Institute.

NTP (National Toxicology Program). 1986. Toxicology and carcinogenesis studies of bromodichloromethane (CAS No. 75–27-4) in F344/N rats and B6C3Fl mice (gavage studies). Precamera draft, National Toxicology Program, Technical Report Series No. 321.

Nestmann, E.R., I. Chu, D.J. Kowbel, T.I. Matula. 1980. Short-lived mutagen in *Salmonella* produced by reaction of trichloroacetic acid and dimethyl sulfoxide. *Can. J. Genet. Cytol.* 22:35–40.

Norwood, D., G. Thompson, J. Johnson, R. Christman. 1985. Monitoring trichloroacetic acid in municipal drinking water. *In:* Jolley, R.L. et al. (eds.), Water chlorination. vol. 5. Chelsea, Michigan: Lewis Publishers, Inc.

Orme, J., D.H. Taylor, R.D. Laurie, R.J. Bull. 1985. Effects of chlorine dioxide on thyroid function in neonatal rats. *J. Toxicol. Environ. Hlth.* 15:315–322.

Parnell, M.J., L.D. Koller, J.H. Exon, J.M. Arnzen. 1986. Trichloroacetic acid effects on rat liver peroxisomes and enzyme-altered foci. *Environ. Hlth. Persp.* 69:73–79.

Rapson, W.H., M.A. Nazar, V.V. Butsky. 1980. Mutagenicity produced by aqueous chlorination of organic compounds. *Bull. Environ. Contam. Toxicol.* 23:590–596.

Rasanen, L, M.L. Hattula, A.U. Arstila. 1977. The mutagenicity of MCPA and its soil metabolites, chlorinated phenols, catechols, and some widely used slimicides in Finland. *Bull. Environ. Contam. Toxicol.* 18:565–571.

Reding, R., P. Fair, C. Shipp, H. Brass. 1986. Measurement of dihaloacetonitriles and chloropicrin in drinking water. Cincinnati, Ohio: U.S. EPA Office of Drinking Water, Technical Support Division.

Regli, S. 1987. Personal communication to Dr. Gerba. Washington, D.C.: U.S. EPA Office of Drinking Water, Criteria and Standards Division.

Rendtorff, R.C. 1954. The experimental transmission of human intestinal protozoan parasites. II. *Giardia lamblia* cysts given in capsules. *Am. J. Hygiene* 59:204–220.

Revis, N.W., P. McCauley, R. Bull, G. Holdsworth. 1986. Relationship of drinking water disinfectants to plasma cholesterol and thyroid hormone levels in experimental studies. *Proc. Natl. Acad. Sci. USA* 83:1485–1489.

Rice, R.G. and M. Gomez-Taylor. 1986. Occurrence of by-products of strong oxidants reacting with drinking water contaminants—Scope of the problem. *Environ. Hlth. Perspect.* 69:31–44.

Rijhsingham, K.S., C. Abrahams, M.A. Swerdlow, K.V.N. Rao, T. Ghose. 1986. Induction of neoplastic lesions in the livers of C57BL x C3HFl mice by chloral hydrate. *Cancer Det. Pre.* 9:279–288.

Robinson, M., R.D. Laurie, R.J. Bull. 1986. Carcinogenic activity associated with chlorinated acetones and acroleins in the mouse skin assay. *Toxicologist* 6:942.

Roe, F.J.C., A.K. Palmer, A.N. Worden, N.J. Van Abbe. 1979. Safety evaluation of toothpaste containing chloroform. 1. Long-term studies in mice. *J. Environ. Pathol. Toxicol.* 2:799–819.

Rook, J.J. 1974. Formation of haloforms during chlorination of natural waters. *Water Treat. Examin.* 23:234–243.

Rosenkrantz, H.S. 1973. Sodium hypochlorite and sodium perborate: Preferential inhibitors of DNA polymerase-deficient bacteria. *Mutat. Res.* 21:171–174.

Schalenkamp, M. 1978. Experience in Switzerland with ozone, particularly in connection with the neutralization of hygenically undesirable elements present in water. Proceedings of 1977 AWWA Annual Conference. Denver, Colorado: American Water Works Association.

Schiff, G.M., G.M. Stefanovic, E.C. Young, D.S. Sander, J.K. Pennekamp, R.L. Ward. 1984. Studies of echovirus-12 in volunteers: determination of minimal infectious dose and the effect of previous infection on infectious dose. *J. Infect. Dis.* 150:858–866.

Shih, K.L., and J. Lederberg. 1976. Chloramine mutagenesis in *Bacillus subtilis. Science* 192:1141–1143.

Simmon, V.F., K. Kauhanen, R.G. Tardiff. 1977. Mutagenic activity of chemicals identified in drinking water. *In: Progr. in Gene. Toxicol.* Scott, D., D.A. Bridges, F.H. Sobels (eds.). New York: Elsevier North-Holland Biomedical Press. pp. 249–258.

Stacpoole, P.W., G.W. Moore, D.M. Kornhauser. 1979. Toxicity of chronic dichloroacetate. *N. Engl. J. Med.* 300:372.

Stevens, A., L. Moore, C. Slocum, B. Smith, D. Seeger, J. Ireland. 1987. By-products of chlorination at ten operating utilities. Presented at the 6th Conference on Water Chlorination. Oak Ridge, Tennessee.

Symons, J., A. Stevens, R. Clark, E. Geldrich, 0. Love, J. DeMarco. 1981. Treatment techniques for controlling trihalomethanes in drinking water. U.S. EPA Report no. 600/2–81–156. Cincinnati, Ohio: U.S. EPA Municipal Environmental Research Laboratory.

Taylor, D.H. and R.J. Pfohl. 1985. Effects of chlorine dioxide on neurobehavioral development of rats. *In:* Jolley, R.L. et al. (eds.), Water chlorination: chemistry, environmental impact and health effects, vol. 5. Chelsea, Michigan: Lewis Publishers, Inc.

Theiss, J.C., G.D. Stoner, M.B. Shimkin, E.K. Weisburger. 1977. Test for carcinogenicity of organic contaminants of United States drinking waters by pulmonary tumor response in strain A mice. *Cancer Res.* 37:2717–2720.

Trussell, A. 1985. *In:* Montgomery, J., Consulting Engineers, Water Treatment Principles and Design. New York: Wiley Interscience.

Trussell, R. and P. Kreft. 1984. Proceedings of AWWA Seminar on Chloramines. Denver, Colorado: American Water Works Association.

Tumasonis, C.F., D.N. McMartin, B. Bush. 1985. Lifetime toxicity of chloroform and bromodichloromethane when administered over a lifetime in rats. *Ecotoxicol. Environ. Safety* 9:233–240.

U.S. Public Health Service. 1963. Inventory of municipal water supplies, U.S. PHS Publication no. 1039. Washington, D.C.

Van Duuren, B.L., S. Melchionne, I. Seidman, M.A. Pereria. 1986. Chronic bioassays of chlorinated humic acids in B6C3Fl mice. *Environ. Health Persp.* 69:109–117.

Ward, R.L. and E.W. Akin. 1984. Minimum infectious dose of animal viruses. *C.R.C. Crit. Rev. Environ. Control* 4:297–310.

Waskell, L. 1978. A study of the mutagenicity of anesthetics and their metabolites. *Mutat. Res.* 57:141–153.

Wlodkowski, T.J. and H.S. Rosenkrantz. 1975. Mutagenicity of sodium hypochlorite for *Salmonella typhimurium. Mutat. Res.* 31:39–42.

Young, T.B., D.A. Wolf, M.S. Kandrek. 1987. Case-control study of colon cancer and drinking water trihalomethanes in Wisconsin. *Intl. J. Epidemiol.* vol. 16, no. 2.

4

Coagulation, Flocculation, Sedimentation, and Filtration

SUMMARY

Coagulation, flocculation, sedimentation, and filtration are commonly used physical and chemical processes that remove materials suspended in water. These processes are highly effective in removing many potentially harmful water contaminants, including suspended sediments, microorganisms, and inorganics.

The chemicals used in these processes—coagulants, coagulant aids, and filtration aids—are of very low toxicity and are used at low concentrations. Analytical methods for determining the exact levels of many of these substances in finished drinking water are not available. However, the many inorganic chemicals resulting from treatment do not appear to pose a threat to public health at the concentrations currently used. The process chemicals may contain potentially toxic impurities which may become part of the finished water.

Data on other treatment chemicals, such as synthetic polymers, are limited. However, the existing data do not indicate that these chemicals pose a significant health risk.

4.1 Process Description

Coagulation, flocculation, sedimentation, and filtration followed by chlorination is the most commonly used series of municipal wastewater treatment processes in the United States.

Coagulation involves the addition of chemicals to alter the physical state of dissolved and suspended solids to facilitate their removal by sedimentation and filtration. The most common primary coagulants are alum $[Al_2(SO_4)_3 14H_2O]$, ferric sulfate $[Fe_2(SO_4)_3]$, and ferric chloride $(FeCl_3)$; doses of each are usually less than 60 mg/L. Additional chemicals that may be added to enhance coagulation include activated silica, a complex silicate made from sodium silicate (Na_2SiO_3) (dose less than 10 mg/L), and charged organic molecules called polyelectrolytes, which include large-molecular-weight polyacrylamides (dose less than 1.0 mg/L), dimethyldiallylammonium chloride (dose less than 3 mg/L), polyamines (dose less than 3 mg/L), and starch (dose less than 5 mg/L). These chemicals ensure the aggregation of the suspended solids during the next treatment step—flocculation. Sometimes polyelectrolytes (usually polyacrylamides) are also added, at doses less than 1.0 mg/L, after flocculation and sedimentation as an aid to the filtration step.

Coagulation may also remove dissolved organic and inorganic compounds. The hydrolyzing metal salts may react with the organic matter to form a precipitate, or they may form aluminum hydroxide or ferric hydroxide floc particles on which the organic molecules adsorb. The organic substances are then removed by sedimentation and filtration, or filtration alone if direct filtration or inline filtration is used (see Chapter 2). Adsorption and precipitation also remove inorganic substances.

Flocculation is a purely physical process in which the treated water is gently stirred to increase interparticle collisions and thus promote the formation of large particles. After adequate flocculation, most of the aggregates will settle out during the 1 to 2 hours of sedimentation.

Sedimentation—another purely physical process—involves the separation from water, by gravitational settling, of suspended particles that are heavier than water. The resulting effluent is then subject to rapid *filtration* to separate out solids that are still suspended in the water. Rapid filters typically consist of 24 to 36 inches of 0.5- to 1-mm-diameter sand and/or anthracite. Particles are removed as water is filtered through the media at rates of 1 to 6 gallons/minute/square foot. Rapid filtration is effective in removing most particles that remain after sedimentation. The substances that are removed by coagulation, sedimentation, and filtration accumulate in sludge. This must be properly disposed of—for example, in well-designed landfills or by other acceptable solid waste management methods.

4.2 Contaminant Removal Efficiencies

Coagulation, flocculation, sedimentation, and filtration effectively remove many contaminants (see Table 4-1). Perhaps most important is the reduction of turbidity. This treatment yields water of good clarity and enhances disin-

Table 4-1. Coagulation, Sedimentation, Filtration: Typical Removal Efficiencies and Effluent Quality

	Coagulation & Sedimentation (% Removal)	Filtration (% Removal)	Filtered Water Concentrations
Total coliform	74–97[a] 60–98[b]	50–98[a] 40–70[b]	<1/100 mL[a] (after disinfection)
Fecal coliform	76–83[a]		<1/100 mL (after disinfection)
Virus	88–95[a] poliovirus and Coxsackievirus	10–98 poliovirus (10^7/L applied)	
Giardia lamblia		97–99.9[d] through coag. sed., and filt.	
Giardia muris	58–99[a]		
Turbidity	40–96[a]		<1 NTU[e]
Trihalomethane formation potential		30–70[c] through coag. sed., and filt.[c]	
Asbestos		99+ through coag. sed., and filt.[f]	$<0.5 \times 10^6$ fibers/L[f]

[a]Berger and Argaman, 1983.
[b]Haas et al., 1985.
[c]Based on TOC removal data summarized by Snoeyink and Chen, 1985.
[d]Al-Ani et al., 1986.
[e]NTU = nephelometric turbidity unit.
[f]McGuire et al., 1983.

fection efficiency. If particles are not removed, they harbor bacteria and make final disinfection more difficult.

Removal of *Giardia lamblia* cysts is also important. These are difficult to kill with disinfectants, and good turbidity removal is required to ensure their removal. Coagulation, sedimentation, and filtration do remove some bacteria and other microorganisms, but disinfection is relied upon to achieve the final removal of these pathogens, especially viruses.

Coagulation, sedimentation, and filtration also remove dissolved inorganic substances. In addition, asbestos fiber concentrations in excess of 10 million fibers/L can be reduced to less than the detection limit of 0.1 to 1 million fibers/L (McGuire et al., 1983). Ions such as lead, trivalent chromium, pentavalent arsenic, and silver can be removed with greater than 80 percent efficiency if they are present in the raw water. By comparison, hexavalent selenium, barium, and hexavalent chromium are poorly removed (efficiency less than 20 percent).

The review by Snoeyink and Chen (1985) documents that some synthetic organic chemicals (SOCs) can be removed by these processes, but that removals are not consistently high. The range of removals for the pesticides DDT, methoxychlor, dieldrin, 2,4,5-T, and endrin was 30 percent to 98 percent, while removals of less than 30 percent were observed for lindane, parathion, aldrin, 2,4-D, rotenone, and toxaphene. Few reports of removals of other compounds were found, although there was evidence that phenol removal was good (60 percent to 80 percent), and that very soluble molecules such as resorcinol, vanillic acid, and citric acid were poorly removed (i.e., less than 10 percent). Unfortunately, neither good data on the residual concentrations of coagulant aids nor analytical methods for detecting these substances are available. A brief study by East Bay Municipal Utility District (EBMUD, 1983) using a colorimetric method to detect residuals of polydiallyl dimethylamide (Catfloc T), showed no detected residual of Catfloc T (i.e., less than 100 μg/L) after its use in direct filtration of a low turbidity, low total dissolved solids (TDS) supply.

4.3 Toxicity of Coagulant Residuals

Use of coagulants and coagulant aids introduces inorganic metal salts (primarily aluminum and iron salts), sulfates, and inorganic and organic polymers into finished drinking water. The health effects of these substances are discussed below. Although the actual health risk cannot be ascertained until precise concentrations of these substances in drinking water have been determined, it appears that extremely high levels of these substances are needed to cause adverse health effects.

4.3.1 Inorganic Metal Salts

Aluminum Salts. Miller et al. (1984) report that aluminum concentrations in raw water are typically less than 16 μg/L and that the median and range of values for surface waters treated with alum are 112 μg/L and 14 to 2,670 μg/L, respectively. The concentration of residual aluminum is especially dependent on the pH of the water applied to the filter and is affected by factors, such as temperature, that influence precipitation kinetics and sedimentation/filtration removal efficiency.

The toxicity of aluminum and some of its salts has been reviewed by the National Academy of Sciences (NAS, 1982) and the World Health Organization (WHO, 1985). In analyses of 1,577 U.S. water samples, Kopp and Kroner (1970) found 456 samples positive for aluminum. Concentrations of soluble aluminum averaged 74 μg/L. Based on this concentration, drinking water contributes approximately 148 μg of aluminum to the dietary intake assuming an average consumption of 2 L water per day. Miller et al. (1984) reported an average dietary contribution from drinking water of 112 μg/L. Compared to the total average dietary intake of 10 to 50 mg/day from all sources, the contribution from water is insignificant—approximately 1 percent.

Aluminum has been associated with some central nervous system toxicity in mammals. However, these effects have only been observed following exposures other than ingestion. Based on an NAS (1982) report, aluminum in drinking water is not believed to present a significant risk:

> Kortus (1967) reported a minimum-effective dose for rats at 1 g/kg/day administered over an 18-day period. Using this value, applying a safety factor of 1,000, and assuming that a 70-kg human consumes 2 L of drinking water daily and that 100 percent of exposure is from water during this period, the 24-hour Suggested-No-Adverse-Response Level (SNARL) was calculated as 35.0 mg/L. This value exceeds the solubility of aluminum in nonacidic solutions. The 7-day SNARL calculated by the National Academy of Sciences is 5.0 mg/L. There are no adequate data from which to calculate a chronic SNARL.

EPA will be proposing a Secondary Maximum Contaminant Level (SMCL) of 50 μg/L for aluminum based on post-precipitation in the distribution system.

Ferric and Ferrous Salts. Iron has a different chemistry than aluminum. Ferric salts are much less soluble than aluminum salts and are preferred for treating both high- and low-pH waters for this reason. Ferric ion can cause esthetic problems in finished waters if its concentration is greater than 0.3 mg/L, but the residual after treatment is usually much less than this level, especially after filtration. Theoretical calculations show that iron should be less than 50 μg/L if pH values are in the range of 4 to 12 (Faust and Aly, 1983).

The health effects of ferric and ferrous salts have been reviewed by the National Academy of Sciences (NAS, 1979) and the World Health Organization (WHO, 1985). Iron is an essential human nutrient. The absorption of iron salts from the gastrointestinal tract has been extensively studied, but the bioavailability of iron in drinking water has not been evaluated.

The EPA Secondary Maximum Contaminant Level for iron is 0.3 mg/L. The Recommended Daily Allowance (RDA) for iron is 10 mg for adult males and 15 mg for adult females (NAS, 1979). Greathouse et al. (1978) reported that the mean concentration of iron in finished water (3,834 samples) was 0.245 mg/L and the maximum concentration was 2.180 mg/L. The WHO recommended a guideline of 0.3 mg/L (at this level, laundry is stained and taste is impaired). As reported by the National Academy of Sciences (NAS, 1980), "Drinking water would contribute approximately 0.5 mg of iron on a daily basis, which is 5 percent of the male requirement and less than 3 percent of the female requirement. For those persons consuming water containing the highest observed value, water would contribute from 17 percent to 44 percent of the daily requirement, depending on sex."

Since precise data on the contribution to the total iron present in finished water by the iron salts used in coagulation are not available, the impact of these agents on health cannot be adequately assessed. However, the levels of iron in finished drinking water are within the beneficial range.

4.3.2 Sulfates

Sulfates occur naturally in water. The levels reported in finished water range from a few tenths of a milligram per liter to several thousand mg/L (NAS, 1977). The only significant adverse effect following ingestion of relatively high doses of sulfates is laxation. Moore (1952) reported that laxation will occur when sulfate plus magnesium ion exceeds 1,000 mg/L.

In a nationwide study of drinking water in 580 cities, Patterson (1981) analyzed water samples from households serviced by municipal water supplies. Sampling was performed in five geographic regions in the United States: the Northeast, Southeast, Midwest, West, and South-Central. In 772 finished systems, the overall mean concentration of sulfate was 101 mg/L. Concentrations ranged from 0 to 820 mg/L for the entire study. The lowest regional mean concentration was reported in the Northeast at 47.5 mg/L. The highest regional mean concentration in the survey was reported in the Midwest at 130 mg/L.

The highest overall mean concentration of sulfate in finished drinking water was recorded in California by Bruvold et al. (1969). The overall mean was 190 mg/L and the range of means was 1 to 1,110 mg/L.

EPA established a Secondary Maximum Contaminant Level of 250 mg sulfate/L of water; the World Health Organization recommends a limit of 400 mg/L. The reported mean concentrations of sulfates in drinking water are well below these levels.

4.3.3 Inorganic Polymers

Activated Silica. Silica, in the form of sodium silicate, is an effective coagulation aid. Natural waters may contain from a few to several thousand milligrams of silicon per liter. The contribution of activated silica to the total silicon content of finished water is not clearly known. The reported median and maximum concentrations of total silica in drinking water are 7.1 and 72 mg/L, respectively (NAS, 1977).

The mechanisms of absorption and excretion of silica have not been thoroughly evaluated. Silicon is an essential trace element for some animals, although its role in human nutrition is unknown. Nothing in the current scientific literature suggests that activated silica is harmful to humans.

4.3.4 Organic Polymers

Natural Polymers. Natural polymers (starches, gums, gelatin) are rarely used in domestic water treatment and human exposure is almost entirely through diet. Their adverse effects on health are essentially nil. Some of these agents are used as sources of nutrition or as food components. The levels ingested through the diet are probably many orders of magnitude more than would possibly be found in drinking water.

Synthetic Polymers. The toxicity of polymers has been evaluated. In general, polymers are not well absorbed from the gastrointestinal tract (although some movement may occur into lymphoid tissue). Polymers are not osmotically active in the gut. Although these compounds are relatively inert biologically, with little effect on health, the monomers such as acrylamide that are present in polymers could elicit toxicity. Detailed information on the monomer content of the polymers used in water treatment and on the bioavailability of these monomers is not available, so no conclusions can be made about their health effects.

Monomers of national concern are being addressed at EPA. For example, EPA will propose regulations for acrylamide in 1988. Concerns about monomers in synthetic polymers have also been addressed through advisory opinions on the acceptability of products for water treatment issued as part of an informal voluntary activity within EPA's Drinking Water Additive programs.

4.4 Filtration Media

Sand and anthracite coal are used for filtration. These materials are inert; they are not consumed during treatment and there are no data suggesting that human exposure occurs from consumption of the finished water. Therefore, it seems unlikely that adverse health effects could be produced by these materials since they are inert substances—i.e., they do not react with the water that is passed through them.

4.5 Risk-Benefit Consideration

The benefits of coagulants, coagulant aids, flocculants, and filtration compounds to water purification are considerable. Their use enables the removal of significant quantities of viruses and other microorganisms, particulates, and some synthetic organics without the apparent addition of significant quantities of the compounds themselves or the formation of any toxicologically important by-products. Data on some of these compounds, such as synthetic polymers, are generally unavailable. Data that do exist for inorganic chemicals resulting from these processes seem to indicate that the health risks are negligible when products of specified quality are used within the prescribed dosing limits. However, more information is needed to more comprehensively assess the overall risks of the chemicals.

REFERENCES

Al-Ani, M.Y., D.W. Hendricks, G.S. Logsdon, G.P. Hibler. 1986. Removing *Giardia* cysts from low turbidity waters by rapid rate filtration. *J. Am. Water Works Assoc.* 78:66.

Berger, P.S., Y. Argaman (eds.). 1983. Assessment of microbiology and turbidity standards for drinking water. EPA Report no. 570–9–83–001. Washington D.C.: U.S. EPA Office of Drinking Water.

Bruvold, W.H., H.J. Ongerth, R.C. Dillehay. 1969. Consumer assessment of mineral taste in domestic water. *J. Am. Water Works Assoc.* 61:575–580.

EBMUD (East Bay Municipal Utility District). 1983. Internal memo from M. Price to K. Carns on "Polymer Residual Study." September 26.

Faust, S.D. and O.M. Aly. 1983. Chemistry of water treatment. Ann Arbor, Michigan: Ann Arbor Science Publishers.

Greathouse, D.G., G.F. Craun, N.S. Ulmer, and A.R. Jharrett. 1978. Relationships of cardiovascular disease and trace elements in drinking water. Presented at the 12th Annual Conference on Trace Substances in Environmental Health, University of Missouri. Columbia, Missouri.

Haas, C.N., B.F. Severin, D. Roy, R.S. Englebrecht, A. Lalchandani. 1985. Removal of new indicators by coagulation and filtration. *J. Am. Water Works Assoc.* 77:67.

Kopp, J. and R. Kroner. 1970. Trace Metals in Waters of the United States: A Five-year Summary of Trace Metals in Rivers and Lakes of the United States (October 1962-September 1967). Cincinnati, Ohio. U.S. Department of the Interior, Federal Water Pollution Control Administration, Division of Pollution Surveillances.

Kortus, J. 1967. The carbohydrate metabolism accompanying intoxication by aluminum salts in the rat. *Experientia* 23:912–913.

McGuire, M.J., A.E. Bowers, D.A. Bowers. 1983. Optimizing large-scale water treatment plants for asbestos fiber removal. *J. Am. Water Works Assoc.* 75:364.

Miller, R.G. et al. 1984. The occurrence of aluminum in drinking water. *J. Am. Water Works Assoc.* 76:84–91.

Moore, E.W. 1952. Physiological effects of the consumption of saline water. Bulletin of Subcommittee on Water Supply, National Research Counsel, January 10, 1952. Appendix B, pp, 221–227.

NAS (National Academy of Sciences). 1982. Drinking water and health, vol. 4. Washington, D.C.: National Academy of Sciences.

NAS (National Academy of Sciences). 1980. Drinking water and health, vol. 3. Washington, D.C.: National Academy of Sciences.

NAS (National Academy of Sciences). 1979. Iron. Report of the subcommittee on iron, National Academy of Sciences Committee on Medical and Biologic Effects of Environmental Pollutants. Baltimore: University Park Press.

NAS (National Academy of Sciences). 1977. Drinking water and health, vol. 1. Washington, D.C.: National Academy of Sciences.

Patterson, J.W. 1981. Corrosion in water distribution systems. Washington, D.C.: U.S. EPA Office of Drinking Water.

Snoeyink, V.L., A.S.C. Chen. 1985. Removal of organic micropollutants by coagulation and adsorption. *The Science of the Total Environment* 47:155. New York, New York: Elsevier/North Holland Biomedical Press.

WHO (World Health Organization). 1985. Guidelines for drinking water quality. Geneva, Switzerland: World Health Organization.

5

Corrosion and pH Control

SUMMARY

Corrosion of water distribution systems releases metallic and nonmetallic substances into the water. It has potential public health as well as economic impacts. Corrosion of this kind can sometimes be reduced by adjusting the pH and carbonate alkalinity of the water. For new or replacement plumbing, the release of corrosion products can also be limited by the judicious selection of materials. Further use of lead solder, flux, and pipes in drinking water applications is banned by the 1986 Safe Drinking Water Act (SDWA) amendments.

The health risks associated with corrosion and pH control center on two factors: (1) the hazards from materials released into drinking water by corrosion and (2) the hazards from use of pH control chemicals. Except for sodium (for persons on low-salt diets), the chemicals added to drinking water for corrosion control or pH adjustment pose essentially no health risks. However, corrosion, which is more of a problem in some regions than others, does release several substances into water, notably lead and, in lower concentrations, cadmium, asbestos, zinc, copper, and iron.

The present drinking water standard for lead is 50 μg/L. This standard is under evaluation. Revised standards for lead were proposed on August 18, 1988. High lead levels are of particular concern for children, who are vulnerable to neurotoxic and behavioral effects of lead at relatively low exposure levels. The drinking water standard for cadmium is 10 μg/L. This standard is also under evaluation. Asbestos has been shown to be a human carcinogen

93

through inhalation exposure, but not by ingestion. EPA is also currently reevaluating the proposed health goals for lead and cadmium; for asbestos, EPA has proposed regulations for fibers larger than 10 micrometers to conservatively protect against the possibility that they may be carcinogenic through ingestion.

5.1 Introduction

This chapter analyzes the health risks associated with methods for controlling corrosion and post-precipitation—two problems that can occur after drinking water treatment. Section 5.2 describes corrosion. Section 5.3 examines the agents released through corrosion—primarily lead, cadmium, and asbestos—and analyzes the risks of noncontrol, i.e., the hazards posed by the substances released through corrosion. Section 5.4 examines the risks of control, i.e., the hazards of the chemical additives used for corrosion and pH control. Both hazards must be considered in assessing the overall health risks of corrosion and pH control. Section 5.5 describes the phenomenon of post-precipitation and the chemical agents involved.

5.2 Corrosion

Corrosion is the wearing away of components of the water distribution system by chemical and/or physical actions that result in the release of metal and nonmetal materials into the water. The deterioration of pumps, pipes, and other parts of water distribution systems—as well as the effect of this deterioration on the consumer's plumbing, hot water heaters, air conditioners, and other devices—has a very serious economic impact on the public and on the utilities themselves. The exact dollar value is very difficult to assess; however, estimates have ranged from a few million dollars to as high as a billion dollars per year.

The health effects of corrosion are caused by the release of both metals and nonmetallic substances into the water. The metals of greatest health concern are lead and cadmium; several other metals (zinc, copper, and iron) are also by-products of corrosion. Asbestos is the nonmetal of greatest concern from the standpoint of potential adverse health effects.

National primary drinking water standards, i.e., Maximum Contaminant Levels (MCLs), exist for both lead (50 μg/L) and cadmium (10 μg/L). Maximum Contaminant Level Goals (MCLGs) have been proposed for lead (zero) (EPA, 1988), cadmium (5 μg/L) (EPA, 1985a), and asbestos (7.1 million fibers longer than 10 μm per liter) (EPA, 1985a). EPA is currently reevaluating these proposed MCLGs. In addition, Secondary Maximum Contaminant Levels (SMCLs) exist for copper (1 mg/L), iron (0.3 mg/L), manganese (0.05 mg/L), and zinc (5 mg/L). SMCLs are non-enforceable standards based on esthetic considerations such as taste and odor.

All metals corrode in contact with water, although the rate varies widely and can be affected by the water quality. Metals react with several substances in the water system, such as hydrogen ions, oxygen, chlorine (and other oxidizing disinfectants such as chlorine dioxide and ozone), as well as other metal ions in solution. Some of these reactions convert the free metal to the ionic, and therefore more soluble, form.

The primary nonmetallic surface that is subject to corrosion is the cement matrix. When asbestos-cement (a-c) pipe is used, this corrosion may release asbestos fibers.

Corrosion control has been a goal of water supply authorities ever since they have been aware of the problem. The approach taken depends on the nature of the substance to be protected. One simple approach is to isolate the metal from the water. To accomplish that goal, pipe can be physically lined with various coatings, such as bitumastic, epoxy paints, and cement-mortar. Other linings can be applied chemically by including a substance in the water that will react to precipitate a coating on the metal surface. Corrosion inhibitors, for example, react with a metal surface to produce a coating that will protect the metal. The most commonly used inhibitors are phosphates and silicates, principally sodium and/or zinc "glassy" (complex) phosphates, sodium silicate, and blends of the two. (Depending on the type used, these chemicals may sequester metal ions in solution.) The doses used range from 0.5 to 5.0 mg/L for phosphates and up to 10 mg/L for the silicates. Their effectiveness varies depending on reactivity with the metal surface of the distribution system; therefore, inhibitor concentrations at the consumer's tap may be equal to those leaving the treatment plant if no reactivity occurs.

To provide a protective coating of relatively insoluble carbonate salts, many utilities control the parameters that affect the presence of carbonate ion. The parameter most commonly controlled is pH, which affects the proportion of alkalinity present as the carbonate ion and thus available to precipitate calcium ions. Generally, when a substantial amount of natural carbonate ion is present in water, adjusting the pH between 8 and 10 can be expected to reduce corrosion. When water is lacking in carbonate ion, then the addition of lime or another source of carbonate ion is required to reduce corrosion.

These treatments affect the quality of the treated water in that they prevent the formation of corrosion by-products. However, they do not remove any of the normal contaminants from water except for a minor amount of calcium ions and carbonate alkalinity through calcium carbonate precipitation. These treatments may add small quantities of sodium, phosphate, silicate, and may affect alkalinity.

5.3 Agents Released Through Corrosion

As discussed above, corrosion releases six primary materials from pipes: lead, zinc, copper, cadmium, iron, and asbestos. Lead is the principal consideration for risk assessment. The health hazards associated with these agents are discussed below. In addition, synthetic organic materials, e.g., certain polymers, are released from plastic pipes and cements used to join them.

One of the most important polymers used in pipes is polyvinyl chloride. There was concern for the diffusion of its residual monomer, vinyl chloride, into drinking water. Recent adoption of strict standards outlined in the National Sanitation Foundation Standard 14, in the American Water Works Association Standard AWWA C900 (AWWA, 1975), and by the American Society for Testing and Materials (ASTM, 1978) concerning the amount of monomer in the finished product has essentially eliminated this problem.

Corrosion is believed to be a greater problem in some regions of the country than in others. In the Northeast and Northwest, corrosive water conditions release lead and other metals into the drinking water. Other areas also probably encounter corrosion problems, but insufficient data have been collected to document the exposure. In any area, residents in homes with corrosive water increase their risk by using "first-draw" or hot water. (First-draw water is water that comes out of the tap in the first few seconds when the tap is turned on after hours of nonuse. When the water has been in the pipes for several hours, it contains higher levels of corrosion by-products. This exposure can be prevented by letting the water run for a few minutes before use.)

5.3.1 Lead

There has been a concerted effort by EPA and others to reduce lead exposure and notify the public of the hazards presented by lead:

- EPA's mandated reduction of lead in gasoline has been associated with a marked reduction in blood lead levels in the United States.
- In 1985, EPA proposed a Recommended Maximum Contaminant Level (now Maximum Contaminant Level Goal) (20 ppb) that was markedly lower than the current lead drinking water standard (i.e., Maximum Contaminant Level of 50 ppb). EPA has subsequently evaluated new data that suggest that lead may be more toxic than previously thought; on August 18, 1988, EPA proposed an MCLG of zero.
- The 1986 Amendments to the SDWA ban all future use of lead solders, flux, and pipes in public water systems and in buildings that

provide water for human consumption after June 19, 1986. States must have a ban in place by June 1988.

* The Food and Drug Administration has banned the use of lead solder in baby food containers.
* In 1987, EPA required public water suppliers to publicly notify persons who may be affected by lead contamination of their drinking water, as specified in the Safe Drinking Water Act.
* In 1987, EPA published "Lead and Your Drinking Water," to alert the public to the hazards associated with lead exposure.
* The recently enacted Lead Contamination and Control Act of 1988 will cause the recall of all water coolers with lead-lined tanks and institutes a testing and remediation program for water coolers in schools.

The following section summarizes data that were used to support the 1985 EPA proposed Maximum Contaminant Level Goal for lead. (Please note that significant additional data have come to light since that proposal; the Agency is now evaluating these new data.)

Exposure. Human populations in the United States are exposed to lead in air, food, water, and dust. In rural areas, Americans not occupationally exposed to lead consume an estimated 40 to 60 μg Pb/day. This level of exposure is referred to as the baseline exposure of the American population to lead in food and ambient air. Forty-four percent of the baseline consumption of lead by children is estimated to result from consumption of 0.1 g of dust per day. Ninety percent of this dust lead is of atmospheric origin (EPA, 1986a).

Adults and older children in the baseline population of the United States receive the largest proportion of lead intake through ingestion or diet. Atmospheric lead may be added to food crops in the field or pasture by atmospheric deposition, or during transportation to the market, processing, and/or kitchen preparation. Metallic lead, mainly from solder on cans or processing equipment, may be added during processing and packaging. Other sources of lead, as yet undetermined, increase the lead content of food as it passes from the field to the table. American children, adult females, and adult males consume an estimated 19, 25, and 36 μg Pb/day, respectively, in milk and nonbeverage foods, plus an additional 7, 11, and 19 μg Pb/day, respectively, in water and other beverages. The added exposure from living in an urban environment is about 28 μg/day for adults, 91 μg/day for children, and 36.2 μg/day for the 6-month-old infant (EPA 1986a; EPA, 1985a).

Lead enters drinking water primarily as a result of corrosion of lead plumbing pipes (interior plumbing, including faucets, lead service lines, and goosenecks) and of lead solder on copper pipes. Corrosion occurs particularly in areas with soft, acidic (pH less than 6.5) waters.

The severity of lead contamination is dependent not only on the corrosivity of source water, but also on the type and age of plumbing materials. Lead levels may vary considerably in different areas of a city, in neighboring houses, or in the same house at different times of the day and during different seasons. Actual human exposures to lead vary with water consumption habits.

All future use of lead solders, flux, and pipes by public water systems and in the plumbing of buildings providing water for human consumption was banned by the 1986 Amendments to the Safe Drinking Water Act (SDWA) as of June 19, 1986. All states were required to enforce the ban by June 19, 1988.

In addition, in 1987 EPA promulgated a public notification requirement for lead. Public water systems are required to identify and provide notice to persons who may be affected by lead in their drinking water, as required by the SDWA. This rule applies to lead contamination resulting from the use of lead in the construction materials of the water distribution system and/or the action of corrosive water (EPA, 1987).

In the case of lead, the corrosivity of water is determined largely by the pH and carbonate alkalinity of the water. Low pHs (below 8.0) and low water carbonate alkalinity (less than 20 mg/L $CaCO_3$) are generally more corrosive toward lead than waters with higher pHs (e.g., 8.5) and higher carbonate alkalinity. However, all water is corrosive to some degree, even water termed noncorrosive or water treated to make it less corrosive. The degree of corrosivity toward lead is the critical factor. The combination of lead service connections or lead solder or galvanized steel pipe and corrosive water is an important source of lead in drinking water, although all water leaches some lead if lead is present in any plumbing materials. Corrosion control will reduce lead in drinking water under any of these conditions.

Health Effects. As part of the regulatory process which led to EPA's proposed Maximum Contaminant Level Goal of zero for lead (EPA, 1988), EPA conducted an extensive review of the scientific literature concerning the adverse health effects of lead in drinking water (e.g., see Figure 5-1). This analysis of the scientific literature has continued as new data have become available. The following is a general discussion of the conclusions EPA reached in 1988 (EPA, 1988; EPA 1986a).

The health effects of lead in both humans and animals are generally measured by relating blood lead (PbB) levels to adverse effects. Numerous studies have correlated PbB levels of people across the United States and adverse health effects. This type of measurement is unusual; for most chemicals the effects are correlated with the intake of the chemical.

While lead affects all individuals at a sufficiently high exposure level,

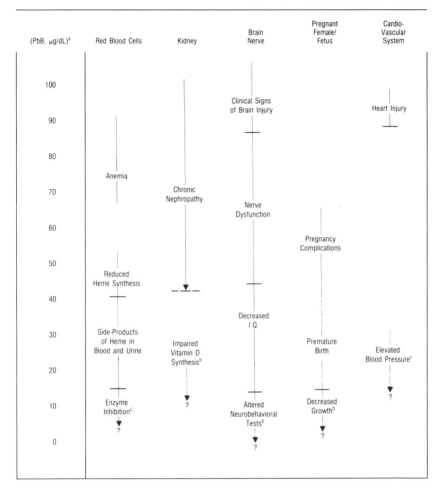

[a]Blood lead levels expressed in μg/deciliters of blood.
[b]Children.
[c]Adults.

Figure 5-1. Adverse health effects of lead.

available data suggest that infants and children are the most sensitive to its toxic effects. The toxic effects produced in children at relatively high blood lead levels (μg/dL) include:

- Encephalopathy or death at approximately 80 to 100 μg/dL.
- Peripheral neuropathies detected in some children at levels as low 40 to 60 μg/dL.

- Chronic nephropathy, indexed by aminoaciduria. This is most evident at high exposure levels (over 100 μg/dL), but may also exist at lower PbB levels (e.g., 70 to 80 μg/dL).
- Colic and other overt gastrointestinal symptoms. These effects clearly occur at high exposure levels and may also occur at lower PbB levels down to 60 μg/dL.
- Frank anemia, evident by 70 μg/dL.
- Reduced hemoglobin synthesis at PbB levels as low as 40 μg/dL, along with other signs of marked heme synthesis inhibition at that exposure level.

Additional studies suggest that other important health effects occur in nonovertly, lead-intoxicated children at lower blood lead levels. Among the most important of these effects are neuropsychological and electrophysiological effects.

Since EPA's 1985 proposal, additional information has become available which suggests that children may be more susceptible to the toxic effects of lead (e.g., cognitive and physical decrements) than EPA had previously thought. This new information includes data associating maternal and fetal blood lead levels with reduced gestational age, lower birth weight, and slowed early postnatal development (both physical and mental) down to 10 to 15 μg/dL blood lead and possibly below (Bellinger et al., 1984; McMichael et al., 1986; Davis and Svendsgaard, 1987). Investigations of postnatal growth and stature also present evidence of a possible negative association in pediatric populations (Schwartz et al., 1986) with blood lead levels ranging down to or somewhat below 10 μg/dL.

The International Agency for Research on Cancer has classified lead as a Group 3 carcinogen, i.e., inadequate evidence for carcinogenicity to humans; sufficient evidence of carcinogenicity to animals (for some salts); inadequate evidence for activity in short-term tests. EPA (1986b) has classified lead as a tentative Group B2 carcinogen, based upon evidence of kidney tumors in orally exposed rats. However, the dosages that induced the kidney tumors in rats were very high (beyond the lethal dose in humans), and several extensive epidemiology studies did not show an association between lead exposure and increased tumor incidence in occupationally exposed workers.

Risk Assessment. It is apparent that corrosion of lead pipe and fixtures contribute to public exposure to lead in drinking water. Thus corrosion may pose a health hazard in some cases. It is anticipated that the 1986 lead ban on solders, flux, and pipes will significantly reduce lead exposure by reducing the number of new lead surfaces. As part of the ongoing regulation of lead in drinking water, EPA is currently examining the health hazards of lead corrosion, as well as ways to mitigate lead corrosion.

5.3.2 Cadmium

The present drinking water standard for cadmium is 10 μg/L. In 1985, EPA proposed an RMCL (now called MCLG) of 5 μg/L for cadmium. The following section summarizes the data EPA used to support that proposal.

Exposure. The major sources of cadmium are diet and cigarettes. The average daily dietary intake of cadmium ranges from 10 to 55 μg/day (Piscator, 1985). Cigarette smoking can contribute an equal amount. Compliance monitoring indicates that 25 public water supplies currently report cadmium levels above 10 μg/L, the current standard. Federal surveys conducted between 1969 and 1980 show that about 27 percent of 707 ground-water supplies had cadmium levels above 2 μg/L; the mean of the positives was about 3 μg/L (EPA, 1985d). In the same surveys, 19.7 percent of 117 surface water supplies had levels above 2 μg/L; the mean of the positives was 3.2 μg/L. None were found to exceed 10 μg/L.

Health Effects. The drinking water standard (MCL) for cadmium is presently 10 μg/L. In 1985, EPA proposed an RMCL (now MCLG) of 5 μg/L. While cadmium can affect virtually all systems in the body at sufficiently high levels of exposure, EPA generally believes that the most sensitive endpoint is renal toxicity. A level of 200 μg of cadmium/gram tissue wet weight of renal cortex in the kidney is believed to be a No-Observed-Adverse-Effect-Level (NOAEL) for renal toxicity. Cadmium has not been shown to be carcinogenic through ingestion. EPA will propose a revised MCL for cadmium in 1988.

Risk Assessment. While high levels of cadmium in drinking water (e.g., 200 μg/L) can result from corrosion, drinking water is generally believed to contribute only slightly to the daily amount of ingested cadmium (i.e., 15 μg/person/day from the diet (FDA, 1986) and, for those who smoke cigarettes, an equivalent amount from smoking). Thus, in general, the total risks presented by the additional exposure to cadmium resulting from corrosion are believed to be minimal.

5.3.3 Asbestos

In 1985, EPA proposed an RMCL (now MCLG) of 7.1 million fibers/L for asbestos. The following section summarizes the data EPA used to support that proposal.

Exposure. Levels of asbestos fibers in drinking water have been summarized by EPA (1980) for 406 cities in 47 states, Puerto Rico, and the District of

Columbia. The distribution of reported asbestos concentrations is shown in Table 5.1.

Table 5-1.   Distribution of Asbestos Fiber Concentrations in Drinking Water for 406 U.S. Cities

Highest Asbestos Concentration, (10^6 fibers/liter)	Number of Cities	Percentage
Below detection limits	117	29
Less than 1	216	53
1 to 10	33	8
Greater than 10	40	10

In 1981, EPA summarized the results of nationwide sampling for asbestos in drinking water from 100 systems. Samples were taken from a representative point in the distribution system of each utility. Levels above detection of 0.08 million fibers per liter (MFL) were found in 12 of the 100 systems. Levels ranged from 0.385 to 1,071 MFL. These and other data from various state studies indicate that asbestos occurs in various drinking water supplies across the country. Asbestos occurs as a result of naturally occurring asbestos in raw water supplies or corrosion of asbestos-cement pipes in distribution systems.

Health Effects. Inhalation studies have shown that various forms of asbestos have produced lung tumors, mesothelioma, and gastrointestinal tract cancer in laboratory animals and humans. The majority of asbestos ingestion studies have failed to produce carcinogenic effects in animals. The National Toxicology Program (NTP) investigated the carcinogenic potential of the ingestion of amosite and tremolite asbestos in rats; the evidence was inadequate to conclude that either was carcinogenic. However, based on studies of ingestion of chrysotile asbestos by F344/N rats, NTP (1984) concluded that "there was some evidence of carcinogenicity" in male rats that were exposed to 1 percent intermediate range (IR) chrysotile asbestos in the diet for the lifetime of the animals.

Several epidemiological studies have investigated potential associations between asbestos fibers in drinking water and gastrointestinal cancer. Marsh (1983) reviewed and evaluated 13 epidemiological studies of ingested asbestos in 5 areas of the United States and Canada for the risk associated with the ingestion of water containing asbestos. He concluded that even though one or more studies found an association between asbestos in water supplies and cancer mortality (or incidence) due to neoplasms of various organs, no individual study or aggregation of studies can be used to establish human risk levels from ingested asbestos.

A report prepared for the Health and Safety Commission of the United Kingdom, as referenced in the *Federal Register* (EPA, 1985e), that examined the available evidence on the health effects of inhaled asbestos concluded that, "In particular, there are no grounds for believing that gastrointestinal cancers in general are peculiarly likely to be caused by asbestos exposure."

EPA's Science Advisory Board (SAB) examined the question of the carcinogenic potential of ingested asbestos in 1984 and concluded that current peer-reviewed evidence for humans and animals does not support the view that asbestos ingested in water causes organ-specific cancers. Subsequently, SAB reexamined the issue and concluded that the data were equivocal and reaffirmed their conclusions summarized above.

In 1985, EPA's Carcinogen Assessment Group (CAG) evaluated the cancer risks of asbestos based on fiber size (EPA, 1985c). CAG concluded that, based on NTP 1985 data, there was an increase in benign polyps of the large intestine for male rats ingesting intermediate-range fibers (greater than 10 μm) at 1 percent of the diet. Although the data indicating potential carcinogenicity of asbestos by ingestion are equivocal, the CAG calculated, based on the one-hit cancer model, that 7.1×10^7, 7.1×10^6, and 7.1×10^5 fibers/L would result in a lifetime excess cancer risk of 10^{-5}, 10^{-6}, and 10^{-7}, respectively. These risk levels are calculated for intermediate-range chrysotile fibers.

Risk Assessment. Asbestos occurs in various drinking water supplies across the country as a result of naturally occurring asbestos in raw water supplies or corrosion of asbestos-cement pipe in the distribution system. Since there is a clear correlation between asbestos exposure via inhalation and carcinogenicity, there is a concern for carcinogenicity via ingestion. After reviewing all the available evidence, EPA very conservatively concluded that asbestos would be considered to have equivocal evidence of carcinogenicity in drinking water and proposed an MCLG at the 10^{-6} risk level (i.e., 7.1×10^6 fibers/L).

5.3.4 Zinc, Copper, and Iron

Zinc, copper, and iron are essential nutrients to humans. Toxicity to humans is evident only at extremely high levels of intake. The levels that humans are exposed to via drinking water are not considered a significant health risk. Chapter 4 discusses the health significance of iron.

Table 5-2. Effects of Corrosion and pH Control on Water Quality

Process	Substance Added	Dose (mg/L)	Residual (mg/L)	Substance Inhibited[a]	Concen. of Substance Inhibited (mg/L)
Corrosion control	Sodium hydroxide	1–20	0.5–10	Pb	0.020–20
				Zn	1–10
				Cu	1–10
				Cd	0.01–0.1
	Calcium hydroxide	1–20	0.5–10		
	Sodium phosphate	1–5	≤4		
			≤4		
	Sodium silicate	1–10	≤3		
	Zinc phosphate	1–5	≤3		
pH control	Sodium hydroxide	1–20	0.5–10		
	Sulfuric acid	1–10	1–10		
	Carbon dioxide	5–50	0–5		
	Calcium hydroxide		0.5–10		

[a]Substances that are generally not present if proper corrosion control is practiced.
Source: Compiled by Edward Singley, Ph.D., Gainesville, Florida.

Table 5-3. Mineral Levels in Biological Materials

Nutrient	Residual[a] Content (mg/L)	Body Content (mg)	RDA (mg)	Intake from Food (mg/day)	Typical Intake (mg)
Calcium	0.5–10	1,000,000	800	500–1,000	5–10
Phosphorus	4	780,000	800	1,500	—
Sodium	0.5–10	80,000	—	1,600–9,600	0.6–3.4
Zinc	3	1,400–2,300	15	12	3.1

[a]Amount found in treated finished water (see Table 5-2).

5.4 Health Hazards of Chemical Additives

The main substances added for corrosion and pH control are sodium and calcium hydroxide, sodium and calcium carbonate, sodium and zinc phosphate, sodium silicate, sulfuric acid, and carbon dioxide (see Table 5-2). The residuals in finished water that result from these additives are calcium, phosphorus, sodium, zinc, and silicates. Toxicological studies on corrosion additives have been conducted (Bull and Craun, 1977) with polyphosphates emphasized. Table 5-3 presents typical residual levels. All these compounds are normal constituents of the body and of foods. These chemicals are believed to present essentially no health risk at the levels found in drinking water. Comparison of the residual concentrations with intake from food or with the Recommended Daily Allowance (Table 5-3) shows that these chemicals present essentially no increased risk at the levels found in treated drinking water. Long-term exposure to sodium in the diet may be a concern for persons who must restrict sodium intake, but drinking water is not a major source of sodium. Sodium is discussed further in Chapter 6.

The silicates are of low toxicity. Acute exposures to high doses produce gastrointestinal effects, but chronic exposures to low levels produce no toxicity. In fact, sodium silicates have been used as a food additive for cattle (Clayton and Clayton, 1981) without any adverse side effects. The removal of toxicologically significant substances (e.g., lead) as shown in Table 5-2 and the introduction of nutritionally beneficial minerals (Table 5-3) appear to far outweigh risks that are incurred from substances introduced by corrosion and pH control. Therefore, these technologies/procedures should be considered generally beneficial practices for treating drinking water.

5.5 Post-Precipitation

The precipitation of any insoluble species after treatment, a phenomenon known as post-precipitation, may cause serious problems such as clogging in the filters, pipes, pumps, and water meters of utility systems as well as in the consumers' plumbing and fixtures. pH control is the most commonly used technique to control post-precipitation. Table 5-2 shows the effects on water quality of corrosion control and pH control.

The two most common post-precipitation problems are caused by the precipitation of either calcium carbonate or aluminum hydroxide. Any post-precipitation of calcium carbonate, such as within the filter medium or in the filter support structures, results in inadequate filtration and thus high turbidity in the finished water. If this problem occurs in the distribution systems, carrying capacity is reduced. Such precipitation is usually controlled either by reducing pH to a value that keeps the carbonate ion concentration below that required to precipitate calcium or by complexing the calcium with 0.5 to 2 mg/L of a complex phosphate, such as sodium hexametaphosphate.

The post-precipitation of aluminum hydroxide results from filtration of alum-treated water at a pH where the aluminum is present in ionic form, i.e., less than about 6 or greater than 8. This precipitation can be prevented by adjusting the pH to 6 to 7, the minimum solubility of aluminum, ahead of filtration.

Two other, less common, precipitation problems are associated with manganese dioxide and magnesium salts. Post-precipitation of magnesium salts, primarily the carbonate or bicarbonate that precipitates when the water is heated, can be prevented by controlling pH and/or the magnesium ion concentration. The effect on water quality of post-precipitation of magnesium salts is minor. On the other hand, post-precipitation of manganese dioxide, caused by inadequate oxidation, results in black residues that may be conveyed to the consumer. Post-precipitation of manganese dioxide can be controlled by adjusting the oxidation step in the treatment process.

REFERENCES

ASTM (American Society for Testing and Materials). 1978. ASTM standard specifications for polyvinyl chloride (PVC) pipe (SDR-PR). ASTM D2241–78. pp. 141–147 in Annual Books of ASTM Standards. Philadelphia, Pennsylvania: American Society for Testing and Materials.

AWWA (American Water Works Association). 1975. Specification 4 in. through 12 in. polyvinyl chloride (PVC) pipe. AWWA C900-75. Denver, Colorado: American Water Works Assoc.

Bellinger, D.C., H.L. Needleman, A. Leviton, C. Waterhaux, M.B. Rabinowitz, and M.L. Nichols. 1984. Early sensory-motor development and prenatal exposure to lead. *Neurobehav. Toxicolog. Teratol.* 6:387–402.

Bull, R.J. and G.F. Craun. 1977. Health effects associated with manganese and manganese-hexametaphosphate complex in drinking water. *J. Am. Water Works Assoc.* 69:662–663.

Clayton, G. and F. Clayton (eds). 1981. Patty's industrial hygiene and toxicology, 3rd edition.

Davis, J.M., and D.J. Svendsgaard. 1987. Lead and child development. *Nature* 329:297–300.

EPA. 1988. Drinking water regulations. Maximum Contaminant Level Goals and National Primary Drinking Water Regulations for Lead and Copper. Proposed Rule. *Fed. Regis.* 53:31516. August 18.

EPA. 1987. National Primary and Secondary Drinking Water Regulations. Public notification requirements for lead. *Fed. Regis.* 52:41534.

EPA. 1986a. Air quality criteria for lead. EPA Report No. 600/8–83/028aF. Research Triangle Park, North Carolina: U.S. EPA Environmental Criteria and Assessment Office.

EPA. 1986b. Guidelines for carcinogen risk assessment. *Fed. Regis.* 51:33992–34003. September 24.

EPA. 1985a. Human exposure to lead. Presented in National Primary Drinking Water Regulations. Proposed Rule. *Fed. Regis.* 50:46968. November 13.

EPA. 1985b. Control of asbestos loss from asbestos-cement pipe in aggressive waters. Bellvue Study: Coop. agreement CR807789010. Project Officer S. Logstion.

EPA. 1985c. Drinking water criteria document for asbestos. 600/X-84/199–1. Cincinnati, Ohio: U.S. EPA Environmental Criteria and Assessment Office.

EPA. 1985d. Human exposure to cadmium. National Primary Drinking Water Regulations. Proposed Rule. *Fed. Regis.* 50:46965. November 13.

EPA. 1985e. Health effects of asbestos. National Primary Drinking Water Regulations. Proposed Rule. *Fed. Regis.* 50:46962. November 13.

EPA. 1980. Ambient water quality criteria for asbestos. EPA Document No. 440/5–80–022. Washington, D.C.: U.S. EPA Carcinogen Assessment Group.

FDA (U.S. Food and Drug Administration). 1986. Memorandum from E. Gunderson, Division of Contaminants, Chemistry Center for Food Safety and Applied Nutrition, Washington, D.C., to Dr. Paul S. Price, U.S. Office of Drinking Water, USEPA, Washington, D.C., November 6, 1986.

McMichael, A.J., G.V. Vimpini, E.F. Robertson, P.A. Baghurst, and P.D. Clark. 1986. The Port Pirie cohort study: Maternal blood lead and pregnancy outcome. *J. Epidemiol. Commun. Health* 40:18–25.

Marsh, G.M. 1983. Critical review of epidemiologic studies related to ingested asbestos. *Environ. Hlth. Perspect.* 53:49–56.

NAS (National Academy of Sciences). 1983. Drinking water and health, vol. 5. National Research Council, Safe Drinking Water Committee. Washington, D.C.: National Academy of Sciences.

NTP (National Toxicology Program). 1984. Board draft. NTP technical report on the toxicology and carcinogenesis studies of chrysotile asbestos in F344/N rats. NTP Technical Report No. 295. Research Triangle Park, North Carolina: National Toxicology Program, Division of Health and Human Services.

Piscator, M. 1985. Dietary exposure to cadmium and health effects: Impact of environmental changes. *Environ. Hlth. Perspect.* 63:127–132.

Schwartz, J., C. Angle, and H. Pitcher. 1986. The relationship between childhood blood lead and stature. *Pediatrics* 77:281–288.

6

Ion Exchange, Lime Softening, and Reverse Osmosis

SUMMARY

Ion exchange, lime softening, and reverse osmosis are commonly used water treatment processes. These technologies are designed to remove specific cations or anions, or groups of them. Most notably, the processes remove hardness ions, predominantly calcium and magnesium. The advantage of these technologies is that they also remove toxic metals and humic acid (a precursor of trihalomethanes [THMs]). The disadvantage, from a public health standpoint, is that two of the ions removed (calcium and magnesium) are essential nutrients. Also, ion exchange may contribute to increased amounts of sodium.

Lime softening can be used to bring the alkalinity and total hardness of the water into better proportion. The process removes or reduces soluble ionic species and some organic contaminants; it adds some impurities that are present in the lime (e.g., iron), as well as any soluble portion of the coagulants used.

Reverse osmosis cleanses water by forcing it through a membrane at high pressure. This process removes most inorganic compounds and high-molecular-weight organic compounds.

Two types of potential health effects may result from these treatment processes: (1) exposure to chemical additives (e.g., acid, salts, and metal ions), and (2) additional sodium intake for sodium-sensitive individuals and others on a low-sodium diet. Based on existing data, it is likely that exposure levels are low for some additives and negligible for others. However, more health effects data are necessary to completely assess the potential health risks.

6.1 Introduction

Three water purification processes—ion exchange, lime softening, and reverse osmosis—are used under various conditions for water treatment. Of these, lime softening is by far the most common; it is used to soften hard waters found in the limestone areas of the midwestern and southeastern United States. Ion exchange is also used to soften these waters, especially in smaller systems in portions of the Midwest. In addition, ion exchange is sometimes used for removing arsenic and nitrate. At present, use of reverse osmosis is limited to certain small brackish water supplies along the East Coast and has also been used in desalination of sea water; however, its success in removing nitrate, radium, uranium, many inorganic substances, and several organic compounds, coupled with technical innovations resulting in reduced cost, will undoubtedly increase the use of this technology.

6.2 Ion Exchange

6.2.1 Process Description

As the name implies, ion exchange involves replacing an ion or group of ions in solution by another ion. The replacing ion is released into solution from a solid substrate, which then attracts the displaced ions from solution.

The process may exchange either cations or anions, depending on the properties of the solid matrix. The cations removed are usually calcium, magnesium, iron, radium, and barium. The anions removed are nitrate and arsenic.

The solid substrates, generally referred to as exchange resins, are predominantly synthetic resins. They are designed to show a strong preference for the ions to be removed from solution and a weak attraction for the ion to be exchanged by the resin. When the resin is saturated with the ions from solution (exhausted), those ions are replaced using a strong solution of the exchangeable ions (regenerated). For removal of calcium and magnesium (softening), radium, barium, and iron, the regenerating solution is sodium chloride (salt). For arsenic, uranium, and nitrate removal, the replacement ions are hydroxide or chloride and thus the regenerant is either hydrochloric acid or sodium chloride. More often, fluoride and arsenic are removed with activated alumina. The alumina is then regenerated with caustic soda and sulfuric acid.

The efficiency of ion-exchange resins depends on their chemical nature, porosity, and mesh size. Stability is also important; if the mesh particles are degraded by physical means, they may contaminate the treated water. Normally, however, the resins do not degrade and do not contribute by-products to the water supply. The resins can be inactivated by oxidation, or can be poisoned by some organic or inorganic compounds. The pores of the resin may be clogged by the colloids in the raw water. The hydroxides of iron, manganese, or aluminum may also precipitate on the surface of the beads of the resins. Proper pretreatment of the raw water may help prevent such deterioration.

6.2.2 Effects on Water Quality

Ion exchange has both positive and negative effects on water quality. While the process removes toxic metals and other ionic contaminants, it adds other ions, principally sodium or aluminum. Table 6-1 shows the effect of various ion exchange processes on water quality. The principal concern is the addition of liberal amounts of sodium, which may present a problem to those on low-sodium diets. Of minor concern is the addition of aluminum (see Section 4.3.1 in Chapter 4).

Table 6-1. Effect of Ion Exchange on Water Quality[a]

		Ion Removed		Ion Added	
		Concentration[b]			
Process	Ion	Untreated	Treated	Ion	Concentration[b]
Softening	Ca	200–500[c]	100	Na^+	46–184
	Mg	80–150[c]	40	Na^+	20–50
	Ra	0.3–50[d]	0–8.5[d]	Na^+	negligible
	Ba	0.5–20	1.0	Na^+	0.1–7
	Fe	0–8	0.1	Na^+	0–14
U removal[e]	$UO_2(CO_3)_x$	0–60[f]	0–5	Cl^-	negligible
NO_3 removal	NO_3–N	20	6–7	Cl^-	14
As removal	AsO_4	1	0.05	Cl^-	negligible

[a]Singley, unpublished data, 1987.
[b]All concentrations listed as mg/L unless otherwise indicated.
[c]mg/L, as $CaCO_3$.
[d]picocuries/liter (pCi/L), sum of [226][Ra] and [228][Ra]. The population-weighted-average concentration is 0.7 to 1.8 pCi/L (EPA, 1986b).
[e]Activated alumina is the exchange medium of choice.
[f]pCi/L. The population-weighted average of the three predominant isotopes is 0.3 to 2.0 pCi/L (EPA, 1986b).

In addition, an unfortunate consequence of ion exchange is that several ion-exchange resins can adsorb natural organic compounds, such as humic acids, that react with chlorine to form THMs. If ion exchange is applied prior to chlorination, it can remove 50 percent to 90 percent of the humic acids. Lower-molecular-weight organic compounds are not removed, however.

6.3 Lime Softening

6.3.1 Process Description

Lime softening is a chemical process to soften raw water. It involves adding lime (calcium hydroxide—$Ca(OH)_2$) or lime and soda ash (sodium carbonate—$NaCO_3$) to the water to precipitate calcium as calcium carbonate ($CaCO_3$) and magnesium as magnesium hydroxide ($Mg[OH]_2$). Acid is added following the precipitation of the calcium, magnesium, and other multi-charged cations so that the pH will fall within an acceptable range.

There are many approaches to lime softening. The most appropriate approach depends on the quality of the raw water. High carbonate hardness waters (i.e., waters with a high ratio of alkalinity to total hardness [carbonate or bicarbonate]) can be softened by using lime alone. When the alkalinity is lower than the total hardness, the difference is referred to as the noncarbonate hardness (NCH). Either soda ash or caustic soda (NaOH) must be used to remove NCH.

Lime softening may also involve addition of an acid to reduce the pH to less than or equal to the pH of calcium carbonate saturation prior to filtration. This is done to prevent post-precipitation (see Chapter 5) in the filters or the distribution system. Post-precipitation can also be prevented by adding sodium hexametaphosphate as a complexing agent. The quantity used varies from 0.5 to 2 mg/L.

6.3.2 Effects on Water Quality

Table 6-2 shows the effects of softening on water quality. The high pH that results from the use of lime, soda ash, or caustic soda also precipitates heavy metal hydroxides or carbonates, such as iron, manganese, radium, and barium. For example, Figure 6-1 shows the relationship between total hardness removal and radium removal. The precipitation also provides a large surface area for adsorption of organic contaminants. For example, an estimated 20 percent to 25 percent of the total organic carbon is removed, depending on the coagulant that is used. The high pH also provides some disinfection, although a primary disinfectant is commonly added, usually after softening.

The only contaminants that could be added by lime softening would be

Table 6-2.   Effect of Softening on Water Quality[a]

Process		Contaminant Removed[b] Typical Concentration (mg/L)		Contaminant Added[c]	Concentration (mg/L)
		Untreated	Treated		
Lime	Ca	200–500[d]	100[d]		
	Mg	80–150[d]	40[d]		
	Fe	0–8	0.1	Al[f,h]	3–5
	Mn	0–5	0.1	$(NaPO_3)_6$[g]	2
	Ra	0.3–50[e]	≤5		
	Ba	0.5–20	<1		
	TOC	0–25	0–20		
	THMFP	50–500	40–400		
Lime-soda	Ca	200–500[d]	100[d]		
	Mg	80–150[d]	40[d]		
	Fe	0–8	0.1	Al[f,g]	3–5
	Mn	0–5	0.1	$(NaPO_3)_6$[g]	2
	Ra	0.3–50[e]	≤5	Na	23–46
	Ba	0.5–20	<1		
	TOC	0–25	0–20		
	THMFP	50–500	40–400		
	NCH	120–170	30		
Caustic soda	TH	200–500	100		
	Ca	200–500[d]	100[d]		
	Mg	80–150[d]	40[d]		
	Fe	0–8	0.1	Na	46–184
	Mn	0–5	0.1	Al[f,g]	3–5
	Ra	0.3–50[e]	≤5	$(NaPO_3)_6$[g]	2
	Ba	0.5–20	<1		
	TOC	0–25	0–20		
	THMFP	50–500	40–400		

[a]Singley, 1987, unpublished data.
[b]Ca = calcium; Mg = magnesium; Fe = iron; Mn = manganese; Ra = radium; Ba = barium; TOC = total organic carbon; THMFP = trihalomethane formation potential; NCH = noncarbonate hardness; TH = total hardness.
[c]Al = aluminum; $(NaPO_3)_6$ = sodium hexametaphosphate; Na = sodium.
[d]mg/L, as calcium carbonate ($caCO_3$).
[e]picocurie/liter (pCi/L).
[f]When alum is used as the coagulant.
[g]May be present at these levels regardless of concentration of other ions.

impurities in the lime and soluble coagulants. Lime impurities are principally heavy metals, such as iron and radium. These are very insoluble at the pH value of softening and are removed along with similar ions present in the untreated water. The organic polymers used as coagulants are strongly absorbed on the voluminous precipitated sludge and are essentially absent in the treated water. The resultant sludge must be properly disposed of—for

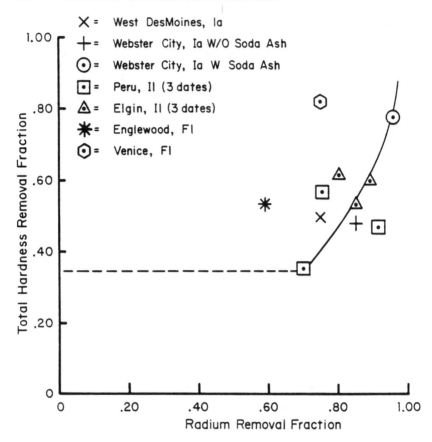

Figure 6-1. Relationship between total hardness removal and radium removal by softening with lime soda. *Source:* Singley et al., 1977.

example, in well-designed landfills or by other acceptable solid waste management methods. When iron or aluminum salts are used as coagulants, they are precipitated along with the other metallic ions. However, a considerable amount of soluble aluminum may pass through the filters and appear in the treated water if the high pH of softening is not reduced prior to filtration. The use of sodium aluminate as a coagulant aid to softening is probably the most important source of aluminum in treated, domestic drinking waters.

6.4 Reverse Osmosis

6.4.1 Process Description

Reverse osmosis is an essentially physical process for cleansing water. Using relatively high pressures—from 100 to 800 pounds per square inch in drinking water plants—the process forces water through a membrane which filters out the larger ions and molecules. The membranes used include cellulose acetate, polyamides, and polysulfones. The principal function of reverse osmosis in municipal water treatment is desalination. Because of the high cost (in excess of $1.00/1,000 gal), it is used for treating sea or brackish waters only when alternative water sources are unavailable.

Small amounts of chemicals are used to prevent scale forming on the semipermeable membrane. Sulfuric acid is usually added to neutralize the carbonate alkalinity, and a complexing agent (usually sodium hexametaphosphate) is added to prevent precipitation of calcium carbonate and magnesium salts. Sodium tripolyphosphate is added to increase water recovery and chlorine is added as a biocide when cellulose membranes are used.

6.4.2 Effects on Water Quality

Most reverse osmosis systems are very effective in removing the majority of inorganic agents and high-molecular-weight organic compounds, including the precursors of THMs. The effectiveness of the process varies with the type of membrane, the pressure used, and the ratio between product and

Table 6-3. Effect of Reverse Osmosis on Water Quality[a]

Contaminant	Initial Concentration	Final Concentration	
Cadmium	400	40	mg/L (as $CaCO_3$)
Magnesium	800	80	" "
Iron	1.0	0.01	" "
Manganese	1.0	0.01	" "
Sulfate	300	30	" "
Chloride	700	70	" "
Radium	25	1.0	pCi/L[c]
THMFP[b]	200	50	mg/L

[a]Taylor et al. (1985).
[b]THMFP = trihalomethane formation potential.
[c]pCi = picocurie(s).

reject water. Typically, reverse osmosis removes about 90 percent to 99 percent of the ions and about 75 percent of the THM formation potential (THMFP). Table 6-3 shows typical values for reverse osmosis treatment of a brackish water. Low-molecular-weight and volatile dissolved organic compounds are only partially removed.

6.5 Potential Health Risks

Ion exchange, lime softening, and reverse osmosis remove substances from drinking water and also add various process chemicals to water. Two types of potential health risks may result from these processes: (1) exposure to the chemical additives, and (2) increased sodium exposure for those on a low sodium diet. The health effects of many of the additives have not been studied, but it is unlikely, based on existing information, that they are present in drinking water at levels that would pose significant health risk. This section examines the health risks of elevated sodium levels and of fluoride, calcium, and magnesium deficiencies.

6.5.1 Sodium

Health risks associated with sodium can be divided into three categories depending on the level of exposure: first, the effects of extreme concentrations in the diet (e.g., hospital accidents that have been fatal to children); second, the effects of chronic exposures to moderate concentrations due to excessive levels in food or excessive use of salt as a condiment (a risk factor for hypertension); and third, the effects of reasonably low concentrations in the diet, including drinking water, on persons who have been placed on restricted salt intake for medical purposes. Sodium levels in the 5 to 20 mg/L range constitute a concern only for those on restricted sodium diets (NAS, 1980).

To assist those who, for medical reasons, require a low-sodium diet, EPA has recommended a guidance level for sodium of 20 mg/L, a level exceeded in most softened water supplies. Reverse osmosis is the only practical way to reduce sodium in water supplies.

Despite much epidemiological data on the subject (Calabrese et al., 1985), there is no compelling evidence that sodium levels up to several hundred mg/L significantly affect the blood pressure of school-age children. However, Colditz and Willet in Calabrese et al. (1985) concluded that "as the effects of drinking water sodium or moderate changes in dietary sodium are likely to be small, objective and uniform measurements of blood pressure and confounding variables, such as dietary sources of sodium and other electrolytes, must be made in future studies. Thus there is need for research to refine the methods of estimating daily sodium intake within a population."

They also noted that "experimental interventions on humans have shown small effects of sodium on blood pressure (0.28 mm Hg/gram sodium). In view of this small response, it is not surprising that many studies with few subjects have inadequate power to detect this apparently small effect."

6.5.2 Fluoride Addition

Ground waters have been reported to contain 0 to 67.2 ppm of fluoride, while most surface waters contain less than 0.1 ppm (Worl et al., 1973). Ion exchange, reverse osmosis, and to some degree lime softening remove excess fluoride from drinking water.

Many communities add fluoride to their drinking water at approximately 1 mg/L to reduce dental cavities. EPA recently established a Maximum Contaminant Level (MCL) and a Maximum Contaminant Level Goal (MCLG), both at 4 mg/L (EPA, 1985; EPA, 1986a). The MCL and MCLG are designed to prevent crippling skeletal fluorosis, a rare disease in this country. Roholm (1937) estimated that a 10- to 20-year daily ingestion of 20 to 80 mg fluoride could result in crippling fluorosis. Only two cases of crippling skeletal fluorosis observed in this country have been linked to fluoride in drinking water. In addition, EPA established a Secondary Maximum Contaminant Level (SMCL—a non-enforceable standard) of 2 mg/L to prevent dental fluorosis, a cosmetic effect (EPA, 1986a). The incidence of objectionable dental fluorosis increases markedly above 2 mg/L.

6.5.3 Calcium and Magnesium

Both calcium and magnesium are essential nutrients and are intimately involved in most body functions. For example, calcium is an essential part of bone and is necessary for the function of the nervous system and numerous other systems. Magnesium is essential to the function of numerous enzyme systems. The amount of magnesium and calcium contributed to the diet via drinking water is typically only a small fraction of the Recommended Daily Allowance (RDA). Although the phenomenon is not totally understood, supplemental intakes of calcium and magnesium in hard waters may help to reduce the risk of cardiovascular disease.

REFERENCES

Calabrese, E.J., R.W. Tuthill, L. Condie (eds). 1985. Inorganics in Drinking Water and Cardiovascular Disease. Princeton, New Jersey: Princeton Scientific Publishing Co., Inc.

EPA. 1986a. National Primary and Secondary Drinking Water Regulations: Fluoride final rule. *Fed. Regis.* 51:11396. April 2.

EPA. 1986b. National Primary Drinking Water Regulations: Radionuclides. Advance notice of proposed rulemaking. *Fed. Regis.* 51:34836. September 30.

EPA. 1985. National Primary Drinking Water Regulations: Fluoride final rule. *Fed. Regis.* 50:47142. Nov. 14.

NAS (National Academy of Sciences). 1980. Recommended dietary allowances, 9th ed. Washington, D.C.: National Academy of Sciences.

Roholm, K. 1937. *In:* A clinical-hygiene study with a review of the literature and some of the experimental investigations. London, England: H.K. Lewis and Co., Ltd. pp. 213–253.

Singley, J.E., B.A. Beaudet, W.E. Bolch, J.F. Palmer. 1977. Costs of radium removal from potable water supplies. EPA Report no. 600/2-77–073. Environmental Protection Technology Series. Cincinnati, Ohio: U.S. EPA, Office of Research and Development, Municipal Environmental Research Laboratory.

Taylor, J.S., D. Thompson, B.R. Snyder, J. Less, L. Mulford. 1985. Cost and performance evaluation of in-plant trihalomethane control techniques. EPA Report no. 600/2–85–138. Cincinnati, Ohio: Water Engineering Research Laboratory.

Worl, R.G., R.E. Van Alstine, D.R. Shaw. 1973. Fluorine. U.S. Geological Survey, Prof. Paper no. 820. pp. 223–235.

Activated Carbon Adsorption and Air Stripping

SUMMARY

Activated carbon adsorption and air stripping are used to control concentrations of organic compounds in water. Activated carbon is a large-surface-area adsorbent that can be used in powdered (PAC) or granular (GAC) form. At present, activated carbon is used primarily to remove organic compounds that give objectionable taste and odor to the water, thus improving the esthetic quality of the water. Increasingly, utilities with contaminated ground waters are using GAC to control the levels of volatile organic chemicals (VOCs) in drinking water. GAC removal efficiencies vary depending on the compound and many other factors, but GAC is highly effective in removing VOCs. PAC is currently not as effective as GAC in removing organics during conventional treatment, but process modifications may lead to increased removal efficiencies in the future.

Air stripping (also called aeration) removes VOCs by transferring them from water to air. It is an effective technology for VOC removal and can be designed and operated to give almost any removal efficiency. It is generally less expensive than GAC. Air stripping may also be used to remove radon from water. EPA has investigated the health risks of air stripping VOCs at selected drinking water treatment facilities and based on that data, concluded that the risks from air stripping were lower than the risks of allowing the VOCs to remain in the drinking water. However, the emissions from air strippers can, and should, be controlled when conditions around the facility warrant concern. EPA is also currently investigating the possible cross-media risks of air stripping radon.

Hundreds, if not thousands, of chemicals may potentially be present in drinking water in trace amounts. According to National Research Council (NRC) data, only 10 percent (by weight) of all the organic compounds in drinking water have been identified, much less assessed. The comparative risk analyses are thus necessarily incomplete, and although activated carbon adsorption and air stripping seem to be effective and relatively safe methods of treating drinking water, more information is needed for definitive assessment of the health risks. The cross-media health concerns related to GAC regeneration and disposal still require extensive investigation.

7.1 Process Descriptions

Activated carbon adsorption and air stripping, sometimes called aeration, are used to control concentrations of organic compounds in water. Both techniques may also be useful for radon removal (Lowry and Lowry, 1987).

7.1.1 Activated Carbon

Activated carbon is a large-surface-area adsorbent that can be used in powdered or granular form. *Powdered activated carbon* (PAC) has a particle size typically less than 100 micrometers in diameter and can be added at several locations in the conventional treatment process train plant (see Chapter 2) to remove trace compounds that cause undesirable taste and odor. Applied this way, PAC generally does not provide good removal of either synthetic organic chemicals (SOCs) or the natural organics that react with chlorine to form trihalomethanes (THMs), but future improvements in the application method may ultimately increase removal efficiency. PAC is removed from the water by either sedimentation or filtration.

Granular activated carbon (GAC) effectively removes many different types of organics (both THMs and SOCs) and radon. In the United States, it is now used primarily to improve the esthetic quality of water by removing organic compounds that cause taste and odor, but GAC can also remove many organic contaminants such as pesticides and other synthetic organic chemicals. GAC is specified in the Safe Drinking Water Act as the standard for best available technology (BAT) for removing organic compounds, and it was designated BAT for seven of the eight volatile organic compounds regulated by EPA on July 8, 1987 (EPA, 1987). With the discovery that volatile organic chemicals (VOCs) contaminate many of our nation's groundwater supplies, utilities have begun to use GAC to control these compounds. In Europe, GAC is commonly used to remove organic substances from surface waters, and its use for this purpose in the United States is likely to increase. GAC removes organics that may appear occasionally because of spills, reduces disinfectant demand, and makes the water less subject to quality changes within the distribution system.

GAC particles range from about 0.3 to 2.5 mm in diameter. They are used to treat water in fixed beds, either as rapid-filter media to remove both particles and dissolved organic compounds, or in a separate contactor located after the rapid filter to remove dissolved organic compounds. The former is called a filter-adsorber and the latter a post-filter adsorber.

7.1.2 Air Stripping

Air stripping usually involves passing water down through a contactor while air is applied upward. VOCs, such as trichloroethylene and tetrachloroethylene, are then transferred from the water to the air because of their volatility; THMs may also be transferred. The volume of air applied per volume of water is an important factor that affects the removal efficiency. Air stripping can also be used to remove radon.

7.2 Removal Efficiencies

7.2.1 Granular Activated Carbon (GAC) Systems

The removal efficiency of GAC systems depends on many factors, including whether a filter-adsorber or post-filter adsorber is used, the quality of the water, the adsorbability of the organic compounds to be removed, the design parameters used for the adsorber, and the frequency of replacement or regeneration of the carbon. Removal efficiencies of adsorbable compounds are usually high when the GAC is fresh, and decrease with time as more water is applied. The length of time of high removal may be short for weakly adsorbed compounds. GAC removal efficiency can be controlled over long periods of time through its design and operation, though this increases the treatment costs.

GAC can remove VOCs from ground water with high efficiency. Typical VOCs include tetrachloroethylene, trichloroethylene, chloroform, carbon tetrachloride, 1,1,1-trichloroethane, benzene, and cis-1,2-dichloroethylene (Love and Eilers, 1982; Snoeyink, 1983). Concentrations of VOCs that have been found in water supplies are given in Table 1-10. Removal efficiencies of 90 percent to 99 percent have been achieved.

The GAC application technique currently used in the United States utilizes shallow GAC beds in which the carbon is replaced every 1 to 5 years (Graese et al., 1987). This technique achieves excellent removal of taste- and odor-causing compounds and 10 percent to 30 percent removal of THM formation potential; however, removal of the THMs or SOCs themselves is negligible.

This is not because of an intrinsically low removal capability of GAC, but because of the way they have been applied during the last 1 to 5 years.

Pilot- and full-scale GAC tests were conducted in Jefferson Parish, Louisiana, to evaluate removal of specific compounds as well as of total organic carbon (TOC) and THM formation potential from surface waters (Lykins et al., 1986). Eighteen volatile and 66 nonvolatile compounds were continuously monitored, but only a few were continuously detected in the influent, and these only at the µg/L or ng/L level. For typical removal rates, see Table 7-1.

Table 7-1.   Typical Removals of Compounds by GAC in Jefferson Parish, Louisiana[a]

Compound	Initial Concentration	Removal (%)	Time Span of Removal (days)
1,2-Dichloroethane	0.1–24 µg/L	100 to 0[b]	90–120
Atrazine	<0.1–5,360 ng/L	100	Removal continuing after 180 days
18 chlorinated insecticides	each <5 ng/L	approx. 100	Removal continuing after 180 days
Total organic carbon	3–4 mg/L	30–70	180

[a]Lykins et al. (1986).
[b]The removals are as high as 100% initially and decrease to 0% with time.

Similar studies were conducted at a treatment plant in Cincinnati, Ohio (EPA, 1984). Many compounds were identified in the influent at the low ng/L level. Compounds such as tetrachloroethylene, 1,2-dichlorobenzene, and hexachloroethane were removed with 85 to 100 percent efficiency over the 300 days of testing. Other compounds, such as diisopropyl ether, benzene, and carbon tetrachloride, were adsorbed with 85 percent to 100 percent efficiency initially, but desorption began after about 160 to 200 days. A few compounds, such as toluene, ethyl benzene, and 1,2,4-trimethylbenzene, were not well adsorbed throughout the run. TOC was reduced from 2.5 mg/L to less than 1 mg/L for about 150 days of operation.

At Thornton, Colorado, TOC at 4.1 to 4.6 mg/L was reduced to 0.6 to 2.1 mg/L over 180 days of operation (Lykins et al., 1986). Lowry and Lowry (1987) have also found that more than 99 percent of radon can be removed using an effective GAC system. For small units, GAC is likely to be cheaper than packed tower aeration for VOC removal, although the latter can achieve high rates of removal.

7.2.2 Air Stripping

Air stripping is highly effective in removing VOCs and many other volatile chemicals, and is generally less expensive than GAC (see Table 2-4). The process can remove over 90 percent of the VOCs in many contaminated ground waters. In fact, air stripping can be designed and operated to give essentially any desired removal, although the cost increases as removal efficiency is increased. For example, air stripping was used to reduce the trichloroethylene concentration from 13,000 to 1 μg/L at one location, and from 30 to less than 5 μg/L at another location (Hess et al., 1983). Certain contaminants, e.g., vinyl chloride and tetrachloroethylene, are easier and therefore less costly to remove than others, e.g., trichloroethylene; carbon tetrachloride is slightly more difficult to remove than trichloroethylene. It should be noted that air stripping is not effective in removing stable halogenated chemicals such as dichloroacetic acid, trichloroacetic acid, and dichlorobutanoic acid. In some locations air pollution control equipment must be used to eliminate the stripped VOCs from the air exiting the process in order to meet air quality regulations. This increases the cost to about the same as for GAC (Crittenden et al., 1987).

7.3 Water Quality Effects of GAC

As discussed in Section 7.2.1, GAC improves water quality by removing organic compounds, including VOCs and THM formation potential. The GAC process may affect water quality in other ways as well.

GAC provides an ideal environment for bacterial growth because it chemically reduces chlorine (which interferes with chlorine's disinfection capability). Bacteria grow on the surface of the GAC and they often slough off and appear in the effluent. However, this biological activity also has an advantage; it removes biodegradable compounds that might cause water quality deterioration in the distribution system. It may also result in the release of biodegradable compounds into the distribution system, especially if ozone is also present. The number of bacteria that actually appear in GAC adsorber effluent depend, therefore, on many factors. Typical results show standard plate counts ranging from 100 to 100,000 colonies per 100 mL and total coliform counts of 1 to 10 colonies per 100 mL. Disinfection sufficient to kill these organisms must follow GAC. However, because GAC removes organic compounds, less disinfectant is generally needed to achieve the desired level of disinfection than is needed without GAC treatment.

In the United States, chlorine is sometimes applied prior to GAC adsorbers to minimize microorganism growth. Laboratory studies by Voudrias et al. (1985) have shown that both hypochlorous acid and monochloramine will promote the polymerization and chlorination of phenols on the GAC surface, resulting in compounds such as polychlorinated, hydroxylated biphenyls, and phenyl ethers. It is not known whether such compounds form under the conditions found in full-scale plants, however. If they do, it may be necessary to stop applying chlorine to GAC adsorbers. Another potential impact of GAC is the introduction of GAC/PAC particulates into water which may have organic contaminants adsorbed to them and which might interfere with disinfection. This impact is normally prevented by backwashing the GAC contactor. Spent GAC must be properly disposed of to mitigate nonwater-quality health hazards.

7.4 Comparative Health Risk Assessment of GAC

7.4.1 Approach

Ideally, a comparative risk assessment of treated and untreated water would consider the concentrations of all chemicals and microorganisms in the water and would determine the extent to which they are increased or decreased by treatment. This is unrealistic, however, because information is lacking in many areas:

- Drinking water may contain hundreds, if not thousands, of chemicals at trace levels; however, many have not been identified. According to NAS (1980) data, only about 10 percent by weight of all the organic compounds in drinking water have been identified. It is likely that many of the remainder are high-molecular-weight, natural, humic-type substances.
- Few data are available on the health effects of the chemicals that have been identified (NAS, 1977).
- The interactions of chemicals in mixtures such as drinking water are unknown.
- The chemical content of drinking water samples may vary substantially from one plant to another and at different times in the same plant.
- Reliable data on GAC removal efficiencies are lacking, and removal efficiencies may vary from one plant to another.

Nevertheless, a limited risk assessment has been conducted using data on GAC performance from a single treatment plant—the Cincinnati Water Works (EPA, 1984). This assessment is highly speculative but is provided to illustrate how a risk assessment would be conducted if it were actually possible to quantify the aggregate risks associated with the various chemicals treated by GAC.

This assessment focuses on contaminant removal by GAC; it also considers the substances that may be added by GAC—primarily microorganisms that result from enhanced microbial activity, and (possibly) organic compounds from interaction of contaminants with GAC itself. Because of variations in treatment performance and limitations in available data, caution must be exercised in generalizing the findings of this assessment.

Data from the Cincinnati Water Works were obtained from Grob closed-loop stripping analyses (CLSA) performed on samples of influent and effluent from one GAC contactor. The influent illustrates the quality of water without GAC treatment, while the effluent illustrates the quality of water with GAC treatment. Influent and effluent were sampled at 4-week intervals over 302 days for a total of 11 samples. Approximately 225 compounds were identified in the influent, with a range of 84 to 130 compounds and an average of 106 on any one date. Except for THMs, all concentrations were in the low ng/L range.

For this assessment, 15 chemicals found in the influent were selected as indicators of drinking water quality (Table 7-2). Twelve are substances for which EPA has established toxicity criteria, i.e., unit cancer risks (UCRs) for carcinogens or Reference Doses (RfDs) for noncarcinogens. The other three were selected to demonstrate the range of treatability with GAC.

Table 7-2. Chemical Indicators of Drinking Water Quality Used for Comparative Assessment

Benzene	Ethylbenzene
Carbon tetrachloride	Hexachloroethane
Chlorobenzene	Styrene
Cyclohexane	Toluene
Dibutyl phthalate	Tetrachloroethylene
Diethyl phthalate	1,1,1-Trichloroethane
Dimethylbenzene	Trichloroethylene
5-Ethyl–1,3-dimethylbenzene	

7.4.2 Hazards of Drinking Water Contaminants Without GAC Treatment

Potency of Noncarcinogens. Table 7-3 presents measures of toxicity for the 15 chemicals. Toxicity for noncarcinogens is expressed in terms of the Reference Dose (RfD)—the total daily dose of a substance that is not expected to cause any adverse noncarcinogenic health effects in humans over a lifetime of exposure. (RfDs were formerly referred to as Acceptable Daily Intakes.) Noncarcinogens are assumed to have thresholds (i.e., exposure levels below which no toxic effects are expected). Therefore, drinking water concentrations resulting in specific chemical exposures below the RfD are unlikely to

Table 7-3.   Toxic Potencies of Selected Chemicals[a]

Chemical	Upper Bound Unit Cancer Risk (mg/kg/day)$^{-1}$	Reference Dose (mg/kg/day)
Benzene	2.9×10^{-2}	NA
Carbon tetrachloride	1.3×10^{-1}	7×10^{-4}
Chlorobenzene	NC	4.3×10^{-2}
Cyclohexane	NC	NA
Dibutyl phthalate	NC	1.0×10^{-1}
Diethyl phthalate	NC	1.3×10^{1}
Dimethylbenzene	NC	NA
5-Ethyl–1,3-dimethylbenzene	NC	NA
Ethylbenzene	NC	9.7×10^{-2}
Hexachloroethane	1.4×10^{-2}	NA
Styrene	3×10^{-2}	2.0×10^{-1}
Toluene	NC	3.0×10^{-1}
Tetrachloroethylene	5.1×10^{-2}	1.4×10^{-2}
1,1,1-Trichloroethane	NC	3.5×10^{-2}
Trichloroethylene	1.1×10^{-2}	7×10^{-3}

[a]NC = Chemical has not been shown to be a potential human carcinogen.
NA = An EPA toxicity criteria is not available for this chemical.

present a hazard to human health (assuming that drinking water is the only source of exposure).

As mentioned above, in the Cincinnati plant, most compounds in the influent to the GAC process were present in the low ng/L range. These low concentrations would result in exposures far smaller than the RfDs in Table 7-3. Approximate margins of safety for the selected noncarcinogenic chemicals are given in Table 7-4. In this table, estimates of average daily exposure (in mg/kg/day) were calculated assuming that a 70-kg adult consumes 2 L/day of drinking water containing each of the noncarcinogens at concentrations of 10 ng/L (a representative concentration in the "low ng/L range"), which is equivalent to 10–5 mg/L. Margins of safety are calculated as the ratio of the RfD to the estimated level of exposure. The procedures used to calculate estimated exposure levels and margins of safety are provided in Table 7-4. Margins of safety in the range of 104 to 107 derived from this analysis suggest that the presence of low levels of noncarcinogens (i.e., in the ng/L range) in drinking water poses virtually no risk to human health.

It should be noted that in the current calculation of margins of safety for noncarcinogens, exposure to an average contaminant concentration was assumed. This assumption would tend to under- or overestimate potential risks

Table 7-4. Illustrative Margins of Safety for Selected Noncarcinogenic Chemicals

Chemical	Estimated Exposure Level (mg/kg/day)[a]	RfD (mg/kg/day)	Margin of Safety (MOS)[b]
Chlorobenzene	3×10^{-7}	4.3×10^{-2}	1×10^{5}
Dibutyl phthalate	3×10^{-7}	1.0×10^{-1}	3×10^{5}
Diethyl phthalate	3×10^{-7}	1.3×10^{1}	4×10^{7}
Ethylbenzene	3×10^{-7}	9.7×10^{-2}	3×10^{5}
Toluene	3×10^{-7}	3.0×10^{-1}	1×10^{6}
Tetrachloroethylene	3×10^{-7}	1.4×10^{-2}	5×10^{4}
1,1,1-Trichloroethane	3×10^{-7}	3.5×10^{-2}	1×10^{5}

[a]Assuming a 70-kg adult drinks 2 L/day of water containing each of the chemicals at a concentration of 10 ng/L (equivalent to 10^{-5} mg/L), the estimated exposure level would be calculated as follows:

$$\frac{10^{-5} \text{ mg/L} \times 2 \text{ L/day}}{70 \text{ kg}} = 3 \times 10^{-7} \text{ mg/kg/day}$$

[b]The margin of safety (MOS) is defined as the ratio of the RfD to the estimated exposure level. For chlorobenzene, with an RfD of 2.7×10^{-2} mg/kg/day, the MOS would be calculated as follows:

$$\frac{2.7 \times 10^{-2} \text{ mg/kg/day}}{3 \times 10^{-7} \text{ mg/kg/day}} = 9 \times 10^{4}$$

for individuals experiencing exposures at the lower or upper ends of the concentration range.

Furthermore, individuals are not exposed to single chemicals in their drinking water, but rather to mixtures of hundreds of chemicals. The effects of noncarcinogenic chemicals on the same target organ or with the same mechanism of action are assumed to be additive (EPA, 1986a). Thus the actual margins of safety for noncarcinogens may be lower than those given in Table 7-4 due to additive effects.

Potency of Carcinogens. It is conservatively assumed that most chemical carcinogens do not have a threshold, and it is EPA's policy to establish Maximum Contaminant Level Goals (MCLGs) for probable carcinogens at zero as an aspirational goal. Measures of carcinogenic potency (expressed as the unit cancer risk [UCR] in units of mg/kg/day^{-1}) for the selected chemicals considered in the Cincinnati Water Works analysis are presented in Table 7-3. Of the 15 selected chemicals, EPA considers six—benzene, carbon tetrachloride, styrene, hexachloroethane, tetrachloroethylene, and trichloroethylene—to be potential human carcinogens. Estimates of carcinogenic risk for these six chemicals can be derived using assumptions similar to those above for noncarcinogens, i.e., that over a lifetime a 70-kg adult consumes 2 L/day of drinking water containing each of the carcinogenic chemicals at concentrations of 10 ng/L. Risk is calculated as the lifetime average daily dose (LADD) times the UCR. Risk levels for these carcinogens using these assumptions are presented in Table 7-5. They range from 3 x 10^{-9} to 4 x 10^{-8}. Risks for chemicals present at higher concentrations would be proportionally greater. In addition, other carcinogens in drinking water would contribute to the total carcinogenic risk posed by drinking water contaminants.

Table 7-5. Illustrative Upper-Bound Risk Levels or Selected Carcinogenic Chemicals

Chemical	Lifetime Average Daily Dose (LADD) (mg/kg/day)[a]	Unit Cancer Risk (UCR) (mg/kg/day)$^{-1}$	Risk Level[b]
Benzene	3×10^{-7}	2.9×10^{-2}	9×10^{-9}
Carbon tetrachloride	3×10^{-7}	1.3×10^{-1}	4×10^{-8}
Hexachloroethane	3×10^{-7}	1.4×10^{-2}	4×10^{-9}
Tetrachloroethylene	3×10^{-7}	5.1×10^{-2}	2×10^{-8}
Trichloroethylene	3×10^{-7}	1.1×10^{-2}	3×10^{-9}
Styrene	3×10^{-7}	3×10^{-2}	9×10^{-9}

[a]Assuming a 70-kg adult drinks for a lifetime 2 L/day of drinking water containing each of the chemicals at a concentration of 10 ng/L (or 10^{-5} mg/L).
[b]Risk = LADD × UCR.

7.4.3 Hazards of Drinking Water Contaminants with GAC Treatment

GAC effectively removes organic contaminants but it may also introduce organic compounds into treated water (e.g., through polymerization and chlorination of water contaminants on the GAC surface). This assessment considers both removal and addition of organic compounds in drinking water by GAC. It also examines the potential of GAC to introduce microorganisms in the drinking water.

Chemical Removal. For more than 25 percent of the 50 chemicals analyzed in the Cincinnati study, the average removal efficiency during 162 days of GAC operation was greater than 90 percent. Removal rates for the 15 selected chemicals in the influent are presented in Table 7-6. Since removal rates can vary substantially from one chemical to another, these 15 chemicals are not necessarily representative of the overall treatability of organic chemicals by GAC, but they do illustrate the range of removal efficiencies for specific chemicals. For the eight selected noncarcinogens with RfDs (see Table 7-4), removal efficiencies ranged from 44 percent to more than 99 percent. The average removal rate for the five carcinogens was 98 percent or more (see Table 7-5).

Table 7-6.   Removal Rates by GAC for Selected Chemicals[a]

| | Chemical[b] | Percent Removal with GAC[c] | |
		Range	Average
High Adsorption	Benzene	99 to 100	99.8 (4)[d]
	Carbon tetrachloride	93 to 100	97.8 (6)
	Chlorobenzene	92 to 100	98.0 (5)
	Hexachloroethane	98 to 100	99.0 (6)
	Styrene	40 to 100	89.0 (6)
	Tetrachloroethylene	98 to 100	99.7 (6)
	Trichloroethylene	98 to 100	99.2 (5)
Moderate	Chyclohexane	21 to 100	70.3 (3)
Adsorption	5-Ethyl-1,3-dimethylbenzene	0 to 99	72.0 (3)
	1,1,1-Trichloroethane	56 to 98	86.0 (5)
Poor	Diethyl phthalate	29	29.0 (1)
Adsorption	Dimethylbenzene	2 to 98	46.7 (6)
	Ethylbenzene	22 to 99	52.0 (4)
	Toluene	−3 to 100	44.5 (6)

[a]EPA (1984).
[b]Removal rates for dibutyl phthalate were not included because concentrations were reported as "not detected" or "not quantified."
[c]Removal rates are for operation of the GAC process for 162 days. Effluent samples taken after 162 days showed negative removal rates for a number of chemicals, indicating that the GAC was exhausted and that desorption was occurring.
[d]Number of sampling points for which percent removal was reported.

Using these removal efficiencies, the amount of risk reduction achieved by GAC can be calculated. For carcinogens, risk reduction is expressed as a decrease in cancer risk level; for noncarcinogens, it is expressed as an increase in the margins of safety. Table 7-7 shows risk reductions for the selected chemicals based on data from Tables 7–4, 7–5, and 7–6. The risk levels shown in Table 7-7 are based on average removal efficiencies. However, it should be noted that these risks will change as the removal efficiencies vary. The greatest reduction in the risk will be seen when 90 percent to 100 percent removal is achieved.

Chemical Addition. While GAC clearly reduces concentrations of many organic compounds in drinking water, it may also introduce toxic organic chemicals through desorption or formation of chemicals from GAC. In a review of GAC, the National Research Council (NAS, 1980) posed the following questions:

• To what extent does the carbon interact with chemical species on the GAC column and produce chemicals of potential health concern? Does the carbon catalyze reactions or enter into reactions producing new chemicals?

Table 7-7. Reduction of Upper-Bound Risks for Illustrative Chemicals[a]

Chemical	Average Percent Removal	Carcinogens		Noncarcinogens	
		Cancer Risk Level without GAC Treatment	Cancer Risk Level with GAC Treatment	Margin of Safety without GAC Treatment	Margin of Safety with GAC Treatment
Benzene	99.8	9×10^{-9}	2×10^{-11}	—	—
Carbon tetrachloride	97.8	4×10^{-8}	9×10^{-10}	—	—
Chlorobenzene	98.0	—	—	9×10^{4}	5×10^{6}
Cyclohexane	70.3	—	—	—	—
Dibutyl phthalate	—	—	—	3×10^{5}	—
Diethyl phthalate	29.0	—	—	4×10^{7}	6×10^{7}
Dimethylbenzene	46.7	—	—	—	—
5-Ethyl-1,3,-dimethylbenzene	72.0	—	—	—	—
Ethylbenzene	52.0	—	—	3×10^{5}	7×10^{5}
Hexachloroethane	99.0	4×10^{-9}	4×10^{-11}	—	—
Styrene	89.0	9×10^{-9}	1×10^{-10}	—	—
Toluene	44.5	—	—	1×10^{6}	2×10^{6}
Tetrachloroethylene	99.7	2×10^{-8}	6×10^{-11}	7×10^{4}	2×10^{7}
1,1,1-Trichloroethane	86.0	—	—	2×10^{6}	1×10^{7}
Trichloroethylene	99.2	3×10^{-9}	2×10^{-11}	—	—

[a]Derived from data in EPA (1984).

• What is the potential for release of chemicals of concern that are formed on GAC?

NRC concluded that "the limited amount of data on catalysis by activated carbon makes it impossible to provide specific conclusions on the potential significance of toxic organic production." Furthermore, while "chlorine, chlorine dioxide, and ozone react readily with carbon and may react with compounds adsorbed on carbon, there is no evidence to indicate that such reactions, under the conditions that exist in water treatment plants, will or will not produce potentially hazardous compounds" (NAS, 1980).

The NRC conclusions point out the difficulties in assessing the hazards of chemicals added to drinking water by GAC. Data from the Cincinnati study, however, can be used to illustrate potential risk, although these data are certainly not representative of GAC performance on a national level.

In the Cincinnati study (EPA, 1984) two of 50 chemicals for which removal data were reported—decanal and nonanal—were consistently higher in the GAC effluent than in the influent. Removal rates for these substances are presented in Table 7-8. The elevated effluent concentrations could be due to leaching of the contactor liner or bacterial degradation of adsorbed organics.

Table 7-8. Chemicals Added to Drinking Water During GAC Treatment[a]

Chemical	Percent Removal[b]	
	Range	Average
Decanal	−680 to 65	−180 (5)[c]
Nonanal	−1300 to 54	−351 (6)

[a]EPA (1984).
[b]Negative removal rates indicate a greater concentration in the effluent than the influent. As in Table 7-6, removal rates are for operation of the GAC process for 162 days.
[c]Number of sampling points for which percent removal was reported.

The toxicologic properties of decanal and nonanal have not been well characterized; however, the inherent hazards of these two chemicals based on their structure and activity are probably far smaller than those of the many known carcinogens and systemic toxicants removed by GAC.

Decanal and nonanal are unlikely to be the only or even the most common chemicals added to drinking water by GAC. As reported in Section 7.3, laboratory studies have shown that both hypochlorous acid and monochloramine will promote the polymerization and chlorination of phenols on the GAC surface, resulting in compounds such as polychlorinated hydroxylated biphenyls and phenyl ethers. More research is needed to establish whether these and other chemicals generated during GAC treatment significantly increase the health risks of GAC-treated drinking water.

Cross-media chemical introductions from regeneration of GAC and disposal of spent GAC may be of health concern but have not yet been fully investigated by the scientific community.

Microbial Activity. Microorganisms are present on the surface of GAC (NAS, 1980); however, the extent to which microbial activity on GAC reduces or increases health risk has not been determined. The National Research Council (NAS, 1980) identified several issues concerning the health consequences of microbial activity on GAC:

- To what extent does microbial activity interact with adsorption during the removal of organic contaminants?
- To what extent does microbial activity remove specific organics (by biodegradation) that are of toxicological concern?
- What is the impact of microbial colonization of GAC beds on microbial contamination of the treated water?
- What products of microbial activity can be expected in the finished water, and are these of toxicological significance?

Additional data are needed before the net effect of microbial activity in GAC treatment can be assessed.

7.5 Comparative Health Risk Assessment of Air Stripping

7.5.1 Volatile Organic Chemicals (VOCs)

Unlike many other technologies, air stripping can be designed and operated to give essentially any desired removal. Using models of air stripper and adsorber technologies, Crittenden et al. (1987) demonstrated that GAC design and operating parameters could be modified to produce up to a 10,000-fold reduction in the concentration of trichloroethylene (a VOC). Available information show air stripping to be as effective as GAC on several other VOCs, although data on removal efficiencies for many chemicals are not available. The removal of VOCs implies that THMs will also be removed by this process. In addition, radon is effectively removed. Because of limitations in data needed to calculate reductions in chemical concentrations by air stripping, a quantitative health risk assessment for air stripping cannot be performed.

Air stripping is known to be highly efficient at removing VOCs in general, many of which are known carcinogens, so it undoubtedly helps to reduce the health risks associated with drinking water. Transfer of VOCs from water to air might be a health concern depending on many factors, including the proximity of the air stripping facility to human habitation, the potential for treatment plant worker exposure, local air quality, local meteorological conditions, the quantity of water processed daily, and the contamination level.

Table 7-9 shows estimated atmospheric concentrations of trichloroethylene (a VOC) 0.2 to 10 km downwind of a typical air stripping installation. These data can be used to estimate the potential human cancer risk from inhalation exposure to vaporized VOCs.

Table 7-9. Air Dispersion Model Results for Trichloroethylene Emissions from a Typical Packed Tower Installation

	Air Concentration (μg/m³) of Trichloroethylene at Downwind Distance (km)				
	0.2	0.5	1.0	5.0	10.0
South	1.0×10^{-1}	2.2×10^{-2}	6.7×10^{-3}	4.9×10^{-4}	1.7×10^{-4}
West	6.5×10^{-2}	1.4×10^{-2}	4.4×10^{-3}	3.3×10^{-4}	1.1×10^{-4}
North	1.0×10^{-1}	2.2×10^{-2}	6.8×10^{-3}	5.0×10^{-4}	1.7×10^{-7}
East	4.1×10^{-2}	8.6×10^{-3}	2.6×10^{-3}	1.9×10^{-4}	6.4×10^{-5}

Source: EPA (1985).

In Table 7-9, the highest air concentration projected is 0.1 μg/m3 and occurs 0.2 km south of the source. The individual lifetime risk of breathing 20 cubic meters of air per day at this location (assuming a 50 percent air to blood transfer of trichloroethylene) is 1.3 x 10^{-7}. Since the concentration in air decreases rapidly as the distance from the source increases, the individual risk from air exposure also rapidly diminishes. This compares with an estimated lifetime risk of 1.4 x 10^{-5} for an individual drinking 2 liters of water containing 50 μg/L of trichloroethylene daily for 70 years (EPA, 1985). A 1985 study by EPA in support of the VOC regulations determined that the cancer risk resulting from exposure to several VOCs in air from aeration of VOC-contaminated water was lower than that resulting from drinking contaminated water. It was also apparent that, in the cases examined, the amounts of VOCs added to air did not significantly increase the cancer risks from airborne contaminants. Activated carbon can be used to remove VOCs from the off-gases of the air stripper, thereby reducing the transfer of VOCs from water to air. This technique could be used in cases where local conditions create unacceptable levels of air contamination; however, it significantly increases the costs.

7.5.2 Radon

On average, drinking water contributes about 1 percent to 5 percent of the indoor air levels of radon. Most radon in air comes from direct geologic transport. The population-weighted-average concentration of radon in drinking water is 50 to 300 pCi/L. Radon, which is a gas, is volatilized from water during showers, baths, and other activities such as washing clothes and dishes. Thus radon can be inhaled as well as ingested. Existing data suggest that inhalation of radon is more toxic than ingestion. Several studies have found a direct link between exposure to radon and its progeny and lung cancer in human populations (EPA, 1986b).

EPA (1986b) has calculated that exposure to natural radon in drinking water could result in up to 30 to 600 excess lung cancer cases per year in the United States. EPA expects to propose an MCLG of zero for radon in 1988, since it is a known human carcinogen.

The risk of air stripping radon from water is a trade-off involving two media, i.e., water and air. Since air stripping removes radon from water to air, it may present an increased health risk to treatment plant workers in close proximity to the aeration system.

Although a comprehensive risk analysis would include these transfer risks, the incremental cancer risk associated with the introduction of radon into the air in this context is expected to be low compared to the risks from natural background levels already in air. Nevertheless, the Agency plans to model the plumes from aeration treatment facilities to estimate the increase of radon and its decay products from air stripping.

As discussed earlier, granular activated carbon (GAC) effectively removes radon and other naturally occurring radionuclides from drinking water. On the other hand, land disposal of the carbon may be complicated by the retention of radon decay products (Pb-210 in particular) in the carbon bed. The Agency plans to address these issues when it publishes the proposed national primary drinking water standards for radionuclides.

REFERENCES

Crittenden, J., et al. 1987. An evaluation of the technical feasibility of the air stripping solvent recovery process. Houghton, Michigan: Department of Civil Engineering, Michigan Technological University.

EPA. 1987. National Primary Drinking Water Regulations. Final rule for 8 VOCs. *Fed. Regis.* 52:25690. July 8.

EPA. 1986a. Guidelines for the health assessment of chemical mixtures. *Fed. Regis.* 51:34014–34025. September 24.

EPA. 1986b. National Primary Drinking Water Regulations. Proposed rule for Radionuclides. *Fed. Regis.* 51:34836. September 30.

EPA. 1985. National Primary Drinking Water Regulations: Volatile synthetic organic chemicals. Final rule and proposed rule. *Fed. Regis.* 50:46910–46911. November 13.

EPA. 1984. Granular activated carbon for removing nontrihalomethane organics from drinking water. EPA-Report no. 600/2–84–165. PB85–120970. Cincinnati, Ohio: U.S. EPA Municipal Environmental Research Laboratory, Office of Research and Development.

Graese, S.L., V.L. Snoeyink, R.G. Lee. 1987. GAC filter-adsorbers. Report to the American Water Works Association Research Foundation, Denver, Colorado.

Hess, A.F. et al. 1983. Control strategy-aeration treatment technique. *In:* Occurrence and removal of volatile organic chemicals from drinking water. Denver, Colorado: American Water Works Association Research Foundation.

Love, O.T. Jr., R.G. Eilers. 1982. Treatment of drinking water containing trichloroethylene and related industrial solvents. *J. Am. Water Works Assoc.* 74:413.

Lowry, J.D. and S.B. Lowry. 1987. Modeling point-of-entry radon removal by GAC. *J. Am. Water Works Assoc.* 79(10)85–88.

Lykins, B.W. Jr., et al. 1986. Granular activated carbon for removing nontrihalomethane organics from drinking water. Cincinnati, Ohio: U.S. EPA, Municipal Environmental Research Laboratory.

NAS (National Academy of Sciences). 1980. Drinking water and health, vol. 32. Washington, D.C.: National Academy of Sciences, National Research Council.

NAS (National Academy of Sciences). 1977. Drinking water and health, vol. 1, Washington, D.C.: National Academy of Sciences, National Research Council.

Snoeyink, V.L. 1983. Control strategy-adsorption techniques. *In:* Occurrence and removal of volatile organic chemicals from drinking water. Denver, Colorado: American Water Works Association Research Foundation.

Voudrias, E.A., R.A. Larson, V.L. Snoeyink, A.S.C. Chen. 1985. Activated carbon: an oxidant producing hydroxylated PCBs. *In:* R.L. Jolley, et al. (eds.), Water chlorination, vol. 5. Chelsea, Michigan: Lewis Publishers.

Acronyms and Abbreviations

ADI	Acceptable Daily Intake
AGI	acute gastrointestinal illness
APHA	American Public Health Association
AWWA	American Water Works Association
AWWARF	American Water Works Association Research Foundation
BAT	Best Available Technology
CAG	EPA's Carcinogen Assessment Group
Catfloc T	polydiallyl dimethylamide
CDC	Centers for Disease Control
CLSA	closed-loop stripping analyses
CNS	central nervous system
DBCP	dibromochloropropane
dL	deciliter
EPA	U.S. Environmental Protection Agency
GAC	granular activated carbon
gm	gram
HEA	Health Effects Assessment
IARC	International Agency for Research on Cancer
IR	intermediate range
kg	kilogram
km	kilometer
L	liter
LADD	lifetime average daily dose
m	meter
MCL	Maximum Contaminant Level
MCLG	Maximum Contaminant Level Goal
MFL	million fibers/liter

mg	milligram
mL	milliliter
MLE	Maximum Likelihood Estimate
MOS	Margin of Safety
NAS	National Academy of Sciences
NCH	noncarbonate hardness
NCI	National Cancer Institute
ND	not detectable
ng	nanogram
NOAEL	No-Observed-Adverse-Effect Level
NOMS	National Organics Monitoring Survey
NORS	National Organics Reconnaissance Survey
NRC	National Research Council
NTP	National Toxicology Program
NTU	nephelometric turbidity unit
PAC	powdered activated carbon
PbB	blood lead
PCE	tetrachloroethylene
pCI	picocurie
pH scale	A scale ranging from 0 to 14 that represents the alkalinity or acidity of a solution. A neutral solution has a pH of 7. Values less than 7 represent increasing acidity; values greater than 7 represent increasing alkalinity.
ppb	parts per billion
RDA	Recommended Daily Allowance
RfD	Reference Dose
RMCL	Recommended Maximum Contaminant Level
RWS	Rural Water Survey
SAB	Science Advisory Board
SDWA	Safe Drinking Water Act and Amendments
SMCL	Secondary Maximum Contaminant Level
SNARL	Suggested-No-Adverse-Response Level
SOCs	synthetic organic chemicals
TCA	1,1,1-trichloroethane
TCAA	trichloroacetic acid
TCE	trichloroethylene
TDS	total dissolved solids
TH	total hardness
THMs	trihalomethanes
THMFP	trihalomethanes formation potential
TOC	total organic carbon
TOX	total organic halide
TTHMs	total trihalomethanes

WHO	World Health Organization
UCR	unit cancer risk
VOCs	volatile organic chemicals

Index